OXFORD BIOGEOGRAPHY SERIES

Editors: A. Hallam, B. R. Rosen, and T. C. Whitmore

OXFORD BIOGEOGRAPHY SERIES

Editors

A. Hallam, Department of Geological Sciences, University of Birmingham.
B. R. Rosen, Department of Palaeontology, Natural History Museum, London.
T. C. Whitmore, Oxford Forestry Institute, University of Oxford.

In an area of rapid change, this series reflects the impact on biogeographical studies of advanced techniques of data analysis. The subject is being revolutionized by radioisotope dating and pollen analysis, plate tectonics and population models, biochemical genetics and fossil ecology, cladistics and karyology, and spatial classification analyses. For both specialist and non-specialist, the Oxford Biogeography Series will provide dynamic syntheses of the new developments.

1. T. C. Whitmore (ed.): *Wallace's line and plate tectonics*
2. Christopher J. Humphries and Lynne R. Parenti: *Cladistic biogeography*
3. T. C. Whitmore and G. T. Prance (ed.): *Biogeography and Quaternary history in tropical America*
4. T. C. Whitmore (ed.): *Biogeographical evolution of the Malay Archipelago*
5. S. Robert Aiken and Colin H. Leigh: *Vanishing rain forests: the ecological transition in Malaysia*
6. Paul Adam: *Australian rainforests*
7. Wilma George and René Lavocat (ed.): *The Africa—South America connection*
8. Yin Hongfu: *The palaeobiogeography of China*
9. Alan J. Kohn and Frank E. Perron: *Life history and biogeography: patterns in* Conus

Life History and Biogeography
Patterns in *Conus*

ALAN J. KOHN
*Department of Zoology,
University of Washington, Seattle*

and

FRANK E. PERRON
*Formerly Department of Zoology,
University of Washington, Seattle*

CLARENDON PRESS · OXFORD
1994

Oxford University Press, Walton Street, Oxford OX2 6DP
Oxford   New York   Toronto
Delhi   Bombay   Calcutta   Madras   Karachi
Kuala Lumpur   Singapore   Hong Kong   Tokyo
Nairobi   Dar es Salaam   Cape Town
Melbourne   Auckland   Madrid
and associated companies in
Berlin   Ibadan

Oxford is a trade mark of Oxford University Press

Published in the United States
by Oxford University Press Inc., New York

© *Oxford University Press, 1994*

*All rights reserved. No part of this publication may be
reproduced, stored in a retrieval system, or transmitted, in any
form or by any means, without the prior permission in writing of Oxford
University Press. Within the UK, exceptions are allowed in respect of any
fair dealing for the purpose of research or private study, or criticism or
review, as permitted under the Copyright, Designs and Patents Act, 1988, or
in the case of reprographic reproduction in accordance with the terms of
licences issued by the Copyright Licensing Agency. Enquiries concerning
reproduction outside those terms and in other countries should be sent to
the Rights Department, Oxford University Press, at the address above.*

*This book is sold subject to the condition that it shall not,
by way of trade or otherwise, be lent, re-sold, hired out, or otherwise
circulated without the publisher's prior consent in any form of binding
or cover other than that in which it is published and without a similar
condition including this condition being imposed
on the subsequent purchaser.*

A catalogue record for this book is available from the British Library

Library of Congress Cataloging in Publication Data
Kohn, Alan J.
*Life history and biogeography: patterns in Conus/Alan J. Kohn
and Frank E. Perron.—[1st ed.]*
p.   cm.—(Oxford biogeography series; 9)
*1. Conus—Indo-Pacific Region—Dispersal. 2. Conus—Indo-Pacific
Region—Geographical distribution. 3. Conus—Indo-Pacific Region—
Ecology. I. Perron, Frank E. II. Title. III. Series.*
QL430.5.C75K75   1994   594'.32—dc20   93-5661
ISBN 0-19-854080-9

Typeset by Advance Typesetting Ltd, Long Hanborough, Oxford
Printed in Great Britain by
the Bath Press Ltd

PREFACE

In this book we take advantage of a number of unusual properties of the marine snail genus *Conus*, to relate aspects of its reproduction, development, and dispersal ability to the geographic distributions of its species. The most striking of these attributes are:

(1) the immense size of the genus; it probably includes more extant species (about 500) than any other genus of marine invertebrates;
(2) the extensive range of variation in species diversity, geographic distribution, ecology, and developmental mode among its species;
(3) a reasonable body of information about comparative biology, especially of the species inhabiting the tropical Indo-Pacific region.

These past studies on *Conus* have led to a number of results of more general biological significance than just knowledge of one genus of snail: comparative ecological studies have enhanced understanding of how coral reef communities are organized. Studies of shell form, secretion, and resorption have helped to understand how molluscs construct and remodel the shell, an important hallmark of the entire phylum. Also, studies of the peptide venoms *Conus* uses to paralyse and overcome its prey prior to ingestion are producing new insights into the functioning of ion channels in nerve cell membranes that control all transmission of nerve impulses.

Here we use data collected mainly by ourselves to focus on comparative patterns of reproduction, embryonic and larval development, dispersal potential, and distribution patterns of *Conus* in the Indo-Pacific tropics, where about one-half of all the species in the genus occur. We try to allow the reader to become specialist and generalist at the same time. For the former, we provide details of each source brood, for this is necessary to establish the data base. For the latter, we synthesize the relationships among developmental patterns, we assess the relationship between dispersal potential and geographic distribution patterns, and we integrate the results with information on other animal groups, particularly invertebrates with similar life history patterns, such as sea urchins and starfish among the echinoderms, and coral reef fishes, which share the same habitats and some of the same resources with many of the *Conus* species discussed here.

In the following chapters we test two main hypotheses:

(1) the overall pattern of life history variation in the genus *Conus* can be interpreted in terms of differences in egg size;
(2) the key attribute egg size, as a determinant of developmental mode and dispersal ability, also explains observed patterns of geographic distribution in the genus.

In Chapter 1, we introduce the genus with a brief discussion of general biology and ecology.

Chapter 2 then focuses in detail on the reproductive characteristics of the genus and poses working hypotheses on the interrelationships among reproductive parameters.

Chapter 3 comprises the data base required to test the hypotheses developed throughout the book. Data on reproduction, embryonic, and larval development in more than 60 Indo-Pacific species of *Conus* are presented, along with the methods used in the larval rearing studies. This uniquely large body of information facilitates the types of comparative analyses reported in the following chapters.

In Chapter 4, data presented in Chapter 3 and tabulated in detail in Appendix Table 1 are used to determine the relationships between such reproductive and life history parameters as egg size, fecundity, hatching size, adult size, and developmental mode.

Chapter 5 extends the analyses of the previous chapter to test the hypothesized relationship between developmental mode and biogeographic patterns.

For purposes of comparison with the Indo-Pacific region, Chapter 6 summarizes available information on reproductive patterns in species of *Conus* elsewhere in the world.

Finally, Chapter 7 summarizes the results of the statistical analyses carried out in Chapters 4 and 5, and discusses the extent to which the data support our main hypotheses.

The authors collected most of the data between 1956 and 1986. We are also grateful to the late James E. Norton, who donated to us many samples of *Conus* egg masses he collected in the Philippines in the 1960s. The research was supported by a series of NSF grants, most importantly DEB-8117945 and BSR-8700523. Bernadette Holthuis, Ian Loch, and John Taylor provided additional material and information on *Conus* reproductive biology. Richard Strathmann and Stephen Kempf read the entire manuscript and offered constructive criticisms. We are also indebted to Gloria Pearson, for the photographs in Plate 1, figures B and C, to the late Poul E. Winther, who executed the drawings of egg capsules, and to Patrick Clarke and Melodie Tune, who prepared the graphs. For helpful discussion and other assistance, we also thank Hau Dang, Richard Emlet, David Leviten, Larry McEdward, Gustav Paulay, Joseph Pawlik, Cristina Soliz, and Florence Thomas. The preserved specimens from this study have been deposited in the B. P. Bishop Museum, Honolulu, Hawaii.

Seattle   A. J. K.
February 1993   F. E. P.

CONTENTS

Plates section falls between pp. 24 and 25

1 Introduction	1
2 Reproductive biology of *Conus*	4
3 Collections and observations	13
4 Relationships among aspects of reproduction and life history	46
5 Relationships of development and biogeographic patterns	57
6 *Conus* development outside the Indo-Pacific region	68
7 Discussion, synthesis, and conclusions	72
Appendix	87
References	99
Index	105

1 INTRODUCTION

He who sees things grow from the beginning will have the best view of them.
Aristotle

For most groups of marine invertebrates, data on reproduction including egg sizes and metamorphosis sizes are very scattered and incomplete.
Christiansen and Fenchel (1979)

Developmental modes among marine invertebrates range from the production of a few large offspring brooded by a protective mother, to the broadcast spawning of millions of tiny planktonic larvae that must fend for themselves as currents disperse them over a vast ocean. Both the existence of such dispersal capabilities and the dramatic differences between even similar taxa contrast conspicuously with most land and freshwater animals, and they must profoundly impact broad patterns of life history, geographic distribution, species longevity, and speciation rates.

One way to approach questions of how this remarkable ontogenetic diversity affects other biological parameters, and how it evolved, is to compare developmental patterns among closely related species of a large taxon. Here we focus on *Conus*, a remarkably diverse, predominantly tropical genus of prosobranch gastropod molluscs with many remarkable attributes. With some 500 species and a geologic history extending back about 55 million years, it has evolved rapidly to become the largest genus in its order and is most likely the largest of marine invertebrates.

About half of these species inhabit the Indo-Pacific region that spans the central tropical Pacific and Indian oceans to the Red Sea. Several species appear to maintain genetic continuity over this entire, remarkably broad geographic range—about one-fourth of the world's ocean area—while others are more restricted, some to a single archipelago or continental coast.

Many *Conus* species are ecologically important, highly specialized predators, especially in coral reef ecosystems. Their feeding specializations consist of a highly derived radular tooth that functions by hypodermically injecting an unusually potent venom. This paralyses the prey, which is engulfed and swallowed whole. The diet is often restricted to only a few prey species, usually all belonging to a single taxonomic class.

GENERAL BIOLOGY OF *CONUS*

Most *Conus* species occur in habitats associated with coral reefs, especially subtidal reef platforms, moats, and lagoons, as well as intertidal benches. Others occupy deeper soft-sediment substrata of bays and continental shelves at depths of tens to a few hundred meters. The number of species that occur together in these environments varies remarkably widely. At its geographic distributional limits at higher latitudes, only one species is usually present, regardless of available habitat types (e.g. in California, the Mediterranean, and Easter Island). In the Indo-Pacific tropics, intertidal benches and subtidal sandy bays typically support 3–9 co-occurring congeners. Maximal *Conus* diversity occurs on subtidal coral reef platforms, where 12–27 species may occur in the same square kilometer of reef. These strong habitat-related gradients in species richness provide an unusual opportunity to compare biological attributes, including patterns of reproduction and development, across a large enough number of distinct but similar species to permit meaningful statistical analyses.

Ecologically, species of *Conus* may be characterized as 'biotically competent', as opposed to 'opportunistic' or 'stress-tolerant', in the terminology of Vermeij (1978). That is, they tend to occupy benign environments that are constantly physiologically favourable, to be long-lived, to have well developed competitive and predator-avoidance mechanisms, and to grow rapidly as juveniles. As Vermeij (1978) pointed out for such species in general, their reproductive potentials and dispersal abilities vary widely.

Linnaeus (1758) named *Conus* for the conical shell with depressed or enrolled spire, typical of most species in the genus. In some species, the shell is biconical with a high spire, and in others the last whorl is more cylindrical than conical. The shell aperture is typically long and narrow, as is the animal's foot that extends from it. The operculum, carried on the dorsum of the extended foot, is far too small to seal the aperture when the foot withdraws. It may be a vestigial structure. The siphon is short but prominent and may be conspicuous in the extended living animal. Several species have characteristic patterns of brightly coloured bands on the siphon (Plate 1A). The head region is small as in most gastropods; it consists of the rostrum or proboscis sheath that bears a pair of tentacles with small, subterminal eyes. The true mouth is located at the tip of an extendible, tubular, intraembolic proboscis, which forms the major component of a remarkably specialized weaponry used to capture prey. The proboscis functions rather like a hypodermic syringe, injecting a detachable, hollow, barbed radular tooth into living prey and pumping a mixture of potent neurotoxic peptides (Olivera *et al.* 1990) through the lumen of the tooth into the wound. These peptides block ion channels in nerve and muscle cell membranes, paralysing the prey organism which is then engulfed and swallowed whole. Most *Conus* species prey exclusively on polychaete annelids. Smaller numbers of species prey on other, unsegmented worms (enteropneusts and echiurans), other gastropods, and fishes (Kohn 1983).

Adult *Conus* are typically quiescent during the day, often sheltered under rocks or completely buried in sediment. They are active nocturnally, when feeding typically occurs. They emerge from their shelters and crawl actively on the surface of the substratum, but their locomotion is slower than even the typical snail's pace. Miller (1974) reported a rate on a smooth surface in the laboratory of 2 cm min^{-1}, slower than any other known marine gastropod that crawls by muscular action of the foot. Marked individuals that have been tracked in their natural habitat move, at most, a few meters in a night (Leviten and Kohn 1980).

Many *Conus* species occupy rather definitely circumscribed habitats as adults, such as coral reefs, intertidal benches, and small sandy bays. This, combined with the low vagility just mentioned, severely limits dispersal by adults. Like most benthic marine invertebrates, and in contrast to most terrestrial vertebrates and insects, dispersal over considerable distances depends on the largely passive mobility of planktonic larvae. All stages of the life history after metamorphosis have rather sedentary, benthic habits.

Growth in *Conus* is indeterminate, periodic increments to the shell are not apparent, and longevity is unknown. Growth curves derived from mark-recapture studies of two species on an Australian Great Barrier Reef suggest that *C. miliaris* reaches modal shell length of 30–35 mm in 3–4 years and may live 6 years, and *C. flavidus* reaches mean shell length of 41 mm in 16 years and may live 30 or more years (Frank 1969; Kohn, unpublished data).

In addition to having numerous and widespread species, the genus *Conus* is characterized by striking gradients in developmental mode and other life history patterns (Perron 1981*a*). These range from long-term planktotrophy, in species that produce very large numbers of small eggs, to nonplanktonic lecithotrophy, associated with small numbers of very large eggs. Marked variation also occurs among species with respect to such important life history parameters as reproductive effort, extent of parental care, age at first reproduction, adult body size, growth rate and life span (Perron 1982, 1986). Because these gradients can serve as dependent variables against which hypothesized causes of observed patterns can be tested, the genus is admirably suited to comparative studies.

Theoretical models beginning with those of Vance (1973) have sought to capture the important parameters of these relationships, typically as a 'strategy' or 'trade-off' that optimizes survivorship to metamorphosis. The basic assumptions of the models are that females devote a given amount of material and energy resources to reproduction; that this can be divided into many small eggs (typically developing into feeding, planktonic larvae), or fewer large eggs (typically developing into planktonic or benthic larvae that do not feed prior to metamorphosis); that the former take longer to reach metamorphic competence; and that mortality is proportional to time to metamorphosis. Emlet et al. (1987) review additional assumptions and how these have been incorporated into some more recent revisions of the models.

Here we present the results of a comparative study of life history traits in 62 species of *Conus* from the vast tropical Indo-Pacific (hereafter, IP) marine region, extending from the Tuamotu, Marquesas, and Hawaiian Islands, westward through the tropical Pacific and Indian Ocean basins to the Red Sea and Persian Gulf. We particularly emphasize the number and size of offspring and modes of embryonic and larval development, but we also consider size at maturity, characteristics of spawn masses, and costs of reproductive efforts. These attributes of life history are both amendable to study and likely to be particularly important to fitness. We use the results as a case study to examine the models of reproductive strategies and the hypothesis that dispersal ability is important in determining biogeographic patterns of sedentary benthic marine invertebrates. Our data on the egg masses, developmental stages, and dispersal capability of larvae are based on 243 samples from the 62 species, about 20 per cent of all known IP *Conus* species. For 26 of these species, data on egg masses are presented for the first time. We briefly summarize previously reported information on the remaining 35 species, and we include records of 27 of the latter from additional geographic regions. Finally, we tabulate previously published and new information in the Appendix.

This body of descriptive information is particularly significant in that it provides an unusually extensive data base for the investigation of life history evolution in the marine environment. Recent attempts at understanding these difficult concepts have focused on the formulation of mathematical models (Vance 1973; Strathmann 1974; Christiansen and Fenchel 1979.) These models, though potentially useful as predictive tools, are only as good as the natural history assumptions on which they are based. Large collections of descriptive data on taxa like *Conus* help clarify interrelationships among life history parameters and permit the derivation of models which more closely correspond to the natural world.

2 REPRODUCTIVE BIOLOGY OF *CONUS*

In this chapter we describe reproductive biology from mating through oviposition, embryonic and larval development, and metamorphosis, with emphasis on generalities applicable to the entire genus. In the course of this discussion, we also introduce questions and hypotheses on which we present relevant evidence in the accounts of particular species in Chapter 3 and a synthesis in Chapter 4.

MATING

The sexes are separate and fertilization is internal in *Conus*, as in most neogastropods. Copulation has rarely been observed, but on one occasion both members of a mating pair of *C. coronatus* were oriented in nearly the same direction, as sketched in Fig. 2.1. The penis was extremely elongated, extending around the front of the male's body and under the right margin of the partner's shell. Copulation of *C. tulipa* has been photographed in an aquarium (Plate 1B). The animals were oriented front to back, with the female lying ventral-side-up on the substratum and the male extended on top with his foot over the aperture of the female's shell. The engorged penis can be seen extending over the anterior part of the male's foot toward the female genital aperture. Zehra and Perveen (1991) reported frequently observing mating of *C. coronatus* but did not describe the process. Rolán (1990) listed aquarium observations of 133 copulation events among four species from the Cape Verde Islands, including three heterospecific matings, but likewise gave no description.

A seminal receptacle is present in female *Conus* (Bergh 1895), but we completely lack information on the interval between mating and spawning, how long sperm may be stored, and whether or not multiple matings and mixed paternity in broods may occur.

Most species of *Conus* have typically solitary habits when they are active, although aggregations of inactive individuals are often found in refuges from adverse physical conditions such as heavy wave action and low tide in shallow water species. Occasionally a male is observed in close proximity to a spawning female. In a few species males and females form mating and spawning aggregations at certain seasons, as is known in some other prosobranchs, e.g. the temperate muricid *Nucella lamellosa* (Spight 1974) and the tropical mesogastropod *Strombus luhuanus* (Catterall and Poiner 1983). Aggregations of *C. geographus* and *C. quercinus* are mentioned in Chapter 4. In the latter species, mating and spawning aggregations in Hawaii typically occur at the same site each February.

SPAWNING

The interval between mating and oviposition is not known for any species of *Conus*, and little information is available on reproductive seasonality, except in Hawaii and Pakistan, where observations have been made throughout the year. In Hawaii, the 19 species we observed spawned only between February and August.

Fig. 2.1 Sketch of positions of male and female *Conus coronatus* while mating, viewed from above. Both individuals adhered to the horizontal substratum by the foot, invisible beneath the shell. Shell lengths: male, 21 mm; female, 22 mm. Uliga Island, Majuro Atoll, Marshall Islands. 3 September 1956.

As part of a population study on *C. pennaceus*, Perron (1983) repeatedly observed tagged individuals to determine the natural spawning periodicity of that species in Hawaii. Over a two-year period, 52 individually marked females were observed to spawn, but none spawned more than once during one 3–4-month breeding season. Females of the same species maintained in the laboratory and fed *ad libitum* rarely spawned more than once per year and never did so during the natural breeding season. The few repeat spawnings observed may have been laboratory artifacts resulting from the abnormally high food intake associated with *ad libitum* feeding. Zehra and Perveen (1991) observed spawning by two species in Pakistan only between April and June, in nature and in the laboratory. In the laboratory females of *C. biliosus* spawned twice, and females of *C. coronatus* spawned 2–3 times, during the season. Zehra and Perveen (1991) also fed the adults during this period but did not state the amount of food provided.

The earliest illustration likely to be of an egg mass of *Conus* (Fig. 2.2A) is by the great seventeenth-century naturalist Rumphius (1705: Plate 32, Fig. II), but the text indicates that he was uncertain of its nature. Adams and Adams (1853) briefly described the egg capsules of *C. capitaneus* from the Indian Ocean. Thorson (1940a) first reported on the early developmental stages of *Conus*. He described and illustrated the eggs, 4-cell stage, and veligers of the species now known as *C. dictator*. Fioroni (1966) summarized information available to that date on developmental morphology of prosobranchs and tabulated comparative aspects of development in families and genera including *Conus*, as well as information on egg size and number, length of development, and larval feeding modes.

At spawning the eggs of *Conus* are encased in soft, thin-walled capsules produced in the capsule gland in the pallial oviduct of the female. These are toughened within the ventral pedal gland as in other neogastropods (Ankel 1929; Fretter 1941), then extruded from the gland and affixed to the substratum (Plate 1C). The only published observations of this process in *Conus* appear to be those of Bandel (1976) on the Caribbean species *C. jaspideus*. He observed that after emerging from the female genital aperture, the soft capsule moves over the right side of the foot, either on the surface or within a temporary groove, and into the ventral pedal gland. The capsule then moves up and down within the ventral pedal gland for 2–4 minutes, the motion slows, and the now firm capsule emerges from the pedal gland after 6–17 minutes and is affixed to the substratum. The animal then moves the foot to position the next capsule, which emerges from the genital aperture about 6 minutes later. The entire deposition cycle for a single egg capsule thus requires about 12–23 minutes, and ends with the capsule affixed to the substratum, usually the underside of a rock. The configuration of the ventral pedal gland may determine the characteristic shape and sculpture of the capsules, thus accounting for interspecific differences, but this remains to be investigated.

EGG MASSES

Conus egg capsules are deposited in a cluster usually referred to as an egg mass (Plate 1D), following Thorson (1940a). Three types of egg masses occur in the genus:

Type I. Each capsule is affixed by a basal plate to a hard substratum; the egg mass consists of several short rows of a few capsules each, often with confluent bases (Fig. 2.2B).

Type II. Only a few capsules are attached to the hard substratum; the others are affixed to earlier deposited capsules, so that their basal plates form bridges and the cluster is quite compact (Fig. 2.2C).

Type III. The first several capsules form an anchor in a sand substratum. They are devoid of eggs and provide an anchor to which the subsequent egg-bearing capsules are affixed (Fig. 2.2D).

Egg masses of Types II and III may be more easily dislodged from their substrata and set adrift than those of Type I, because relatively few capsules anchor the entire mass. Type II egg

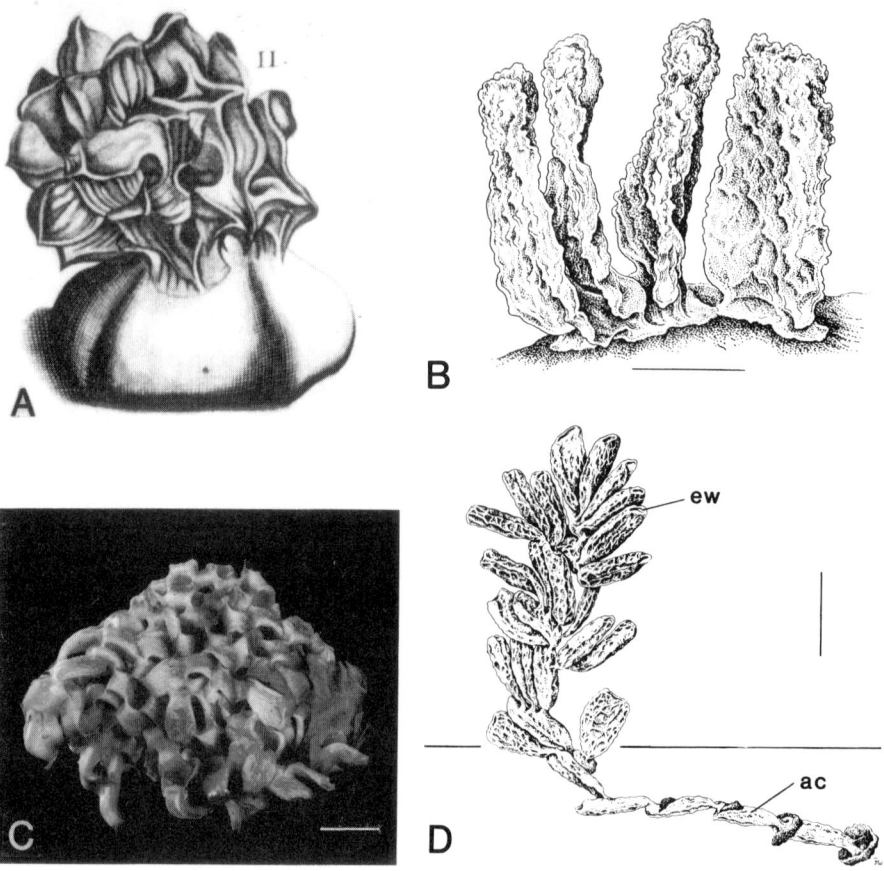

Fig. 2.2 (A) Illustration of an 'Eierstuk' from Rumphius (1705: Plate 32, Fig. II). Probably a *Conus* egg mass attached to the test of a spatangoid sea urchin. (B) Type I *Conus* egg mass; each capsule is affixed to a hard substratum. *Conus litteratus*; Guam. Scale bar = 1 cm. (C) Type II *Conus* egg mass; most capsules are affixed to previously laid capsules. *Conus furvus*; Philippines. Scale bar = 1 cm. (D) Type III *Conus* egg mass; the first five capsules are deposited below the sand surface, lack eggs, and form an anchor for the rest of the egg mass. The horizontal line indicates the water–sand interface. ac, anchor capsule; ew, exit window. *Conus figulinus*; Sri Lanka. Scale bar = 2 cm.

capsules are often thicker and tougher than those of the other types (Perron 1982; Perron and Corpuz 1982). These properties increase the likelihood that if a Type II egg mass becomes dislodged and some of its capsules become punctured and filled with air, the entire mass may float and be carried considerable distances by ocean currents. Such drifting egg masses have been documented (Natarajan 1957; Kohn 1961a) and have been observed several times by the second author. In species of *Conus* lacking planktonic larvae, such an egg mass could provide an alternate means of dispersal.

In *Conus* species that occur on topographically complex coral reefs, the egg capsules, in masses of Type I or Type II, are usually affixed to a protected site in a depression or the underside of a large coral head or boulder, where the capsules are suspended from above and extend down into a layer of actively moving water above the reef substratum. Species that occur on broad expanses of sand bottom and have Type I or II egg masses affix them to algae, sponges, shells, and sometimes to discarded objects of human origin. Type III egg masses, at present known only in one species, *C. figulinus*, are deposited directly within the sand substrate. At present, virtually nothing is known of the mechanism and process of nest site selection, or the degree of specialization by different species to nest sites of different types.

Upon completion of oviposition, the female leaves the egg mass. Brooding, which occurs in many mesogastropods (e.g. *Cypraea, Crepidula*) but not in neogastropods, is clearly not possible in *Conus*; the narrow foot is far smaller than the extent of the egg mass. Apparently there is no guarding or patrolling of the nest either. This is another attribute that *Conus* shares with other neogastropods, in contrast to mesogastropods with large egg masses.

EGG CAPSULES

The individual *Conus* egg capsule is shaped like a flattened vase or flask, usually with convex edges, a very short, broad stalk attaching it to an adhesive basal plate, and a translucent, apical exit window through which the larvae emerge at hatching (Figs 2.2, 3.5). In some species, two ridges project basally from the edges of the exit window on one face (Fig. 3.7); the rest of the surface may be smooth or bear a characteristic or irregular pattern of ridges. These often consist of a peripheral row of simple, evenly, or irregularly spaced ridges directed inward or basally from the edges (Fig. 3.5), or a network of ridges over the entire face of the capsule (Fig. 2.2B, D). Egg capsules of numerous species have been illustrated (Kohn 1961*a,b*; Bandel 1976).

Throughout embryonic development, *Conus* egg capsules remain remarkably free of adherent organisms. After hatching occurs, empty capsules commonly acquire a film of microscopic algae. Viable capsules likely deter epibionts chemically, but the mechanism is completely unknown.

Details of egg capsule microstructure are known only for two *Conus* species, the Atlantic *C. floridanus* and *C. jaspideus* (D'Asaro 1988). Three fibrous layers constitute the capsule wall. This is in contrast to other neogastropod groups; the Muricoidea and Buccinoidea studied by D'Asaro (1988) have four layers. It is possible, but not yet demonstrated, that the thin, outermost layer is applied to the capsule in the ventral pedal gland.

Each *Conus* egg capsule is made up of three major components: ova, capsule wall material, and intracapsular fluid. Caloric analysis of these components in the spawn of *C. pennaceus* shows that the ova are energy rich, with an ash-free caloric content of about 6.2 kcal g^{-1}. The capsule wall material, probably consisting primarily of proteins and mucopolysaccharides, contains 5.5 kcal g^{-1}. Both ova and capsule wall material have relatively low inorganic components, as indicated by their ash contents of approximately 2 per cent and 12 per cent, respectively. In contrast, the dry weight of intracapsular fluid is 75 per cent ash and the fluid has an ash-free caloric content of 4.2 kcal g^{-1}. The caloric content of a typical *C. pennaceus* egg capsule is partitioned as 48 per cent ova, 37 per cent capsule wall material and 15 per cent intracapsular fluid.

The total amount of energy invested in an egg mass can be substantial and may be used to calculate reproductive effort, the proportion of a female's energy budget that is devoted to reproduction. Reproductive effort is an important life history parameter because it represents a measure of the cost of reproduction to the individual female. For example, high levels of reproductive effort may imply increasing physiological stress, vulnerability to predation, or reductions in future fecundity resulting from decreased growth. In *Conus*, reproductive effort, as well as other reproductive and early life history characteristics, varies considerably among species. We briefly indicate the ranges of variation here, present the data on which they are based in Chapter 3, and examine their interrelationships and possible causes in terms of reproductive strategies in Chapter 4. In the case of reproductive effort, the values range from 18 per cent (*C. quercinus*) to 34 per cent (*C. pennaceus*) among the species in which it has been measured.

The fraction of reproductive effort allocated to extraembryonic protective structures (capsule wall material and intracapsular fluid) can also be used as a convenient measure of parental care, another important life history parameter. Neogastropods provide excellent material for the study of parental care since the egg capsule represents the only contribution the female makes toward the survival of her offspring. This is in contrast to the situation in other animal groups in which the presence of brooding, placental nutrition, and complex protective behaviour make quantification of parental care much more difficult. The costs of parental care in *Conus* range from less than 30 per cent to more than 50 per cent of total reproductive energy.

Conus species that have larger eggs and longer intracapsular prehatching development of embryos produce capsules that are more costly than those with smaller eggs and shorter prehatching durations. The proportion of energy allocated to capsule wall material is significantly positively correlated ($r = 0.88$) with egg size and prehatching developmental time. The energy-rich capsules produced by species with larger, slowly developing ova are also thicker and stronger ($r = 0.91$) than those produced by species with small, quickly developing ova (Perron 1981*b*). Perron (1981*b*) suggested that this observed pattern results from the increased likelihood that more slowly developing ova that are retained in the benthic environment for longer periods of time will be discovered by benthic predators. These embryos must therefore be protected in stronger, more energetically expensive capsules.

The proportion of energy allocated to ova compared to extraembryonic protective structures also varies within species as a function of female age or size. In *C. pennaceus*, Perron and Corpuz (1982) showed that larger females produce larger numbers of larger capsules with thicker, stronger walls. These contain more eggs per capsule, but the relationship is non-linear, and their energetic cost per ovum is higher than for smaller, younger females. Thus the costs of parental care increase as females grow and age.

Within species, the pattern of ova packaging in *Conus* may be controlled by simple physical constraints operating at the level of the individual capsule. Since the number of ova per capsule is linearly related to capsule surface area (Perron and Corpuz 1982), the maximum number of eggs a capsule can support is likely to be limited by the amount of respiratory surface area provided by the capsule walls. For capsules of the same shape, the surface:volume ratio decreases with increasing capsule size, thereby reducing net gas transport per unit of capsule contents. Also, the thicker walls of large capsules may inhibit gas transport.

Here we extend these approaches to comparisons among other species of the genus. We examine both intraspecific and interspecific trends in the relationships between size of parent and size of eggs and egg capsules, number of eggs per capsule, and fecundity or number of eggs per egg mass.

EGGS

The eggs of most *Conus* species are spherical. They vary in diameter from about 125 μm to 1 mm, an unusually broad range for a single genus. Because the volume of a sphere scales as the third power of diameter, egg volume, and mass, vary by a much larger factor: the volume of the largest *Conus* eggs is 500 times that of the smallest. Within the IP region, the range in diameter is 125–850 mm, equal to a volume range of more than 300 times. Volume or mass is a better estimate than diameter of the energy commitment per egg, and of the number of eggs that a female's reproductive system can accommodate.

The number of eggs per capsule also ranges unusually widely. *Conus diminutus* from the eastern Atlantic, one of the smallest known species with maximum adult shell length of 15 mm, produces only 4–5 eggs per capsule (Trovão and Rolán 1986). *Conus furvus* produces only 6–12 eggs per capsule, the fewest for any known IP species. At the opposite extreme is the IP species *C. vexillum*, with more than 50 000 eggs per capsule.

DEVELOPMENT OF THE EMBRYO

Ova enter the egg capsules after fertilization but usually prior to the first cleavage division. All eggs in a capsule typically complete embryonic development. Nurse eggs, consumed by siblings within the egg capsule, are present in a number of neogastropod groups but are not known to occur in *Conus*. Within a *Conus* egg capsule, all eggs typically develop synchronously. The time required to complete oviposition is not known; it likely lasts one or more days, as the embryos in some capsules may be one to a few days ahead in their development compared with those in other capsules, presumably deposited later, in the same egg mass.

The limited information we present suggests that developmental patterns are generally consistent in different populations of the same species, even if they are geographically widespread. To our knowledge there has never been a suggestion of poecilogony—different developmental modes in the same species—in *Conus*. This phenomenon has been proposed to occur in a number of other prosobranch gastropod taxa, but it is now generally discredited (Bouchet 1989; Hoagland and Robertson 1988).

Earlier studies (Kohn 1961a,b) described the time course of embryonic development of nine *Conus* species. The range in number of days after spawning to reach each developmental stage in these species is: early cleavages (to 4- and 8-cell stages) 0–1; blastula: 1–5; gastrula: 4–5; early trochophore-equivalent: 4–9; early veliger: 6–13; well-developed veliger: 8–15; hatching veliger: 14–15; hatching veliconcha: 16. Here we report new information on prehatching development for 10 additional IP species.

Perron (1981a) reported the prehatching period of 18 species from Hawaii to range from 11 to 26 days ($x = 14.8$; SD $= 3.4$). He found significant correlations between prehatching period and egg size (direct; $r = 0.97$), size at hatching and egg size (direct; $r = 0.98$), duration of minimum planktonic period and egg size (inverse; $r = 0.87$), and duration of minimum planktonic period and hatching size (inverse; $r = 0.84$). These analyses strongly implicate egg size as a critical determinant of embryonic and larval development and of life history patterns.

LARVAL BIOLOGY, DEVELOPMENT, AND GROWTH

Except for a few species in which hatching occurs during the veliconcha stage and the planktonic phase is absent or lasts only a few hours, hatching in *Conus* releases veliger larvae from the egg capsule (Kohn 1961a,b; Perron 1981a). These gastropods thus lead a dual life-style that is typical of many marine invertebrates—the egg develops into a small, planktonic larval stage that disperses at the whims of ocean currents (Plate 2A,B), and the larva metamorphoses into a slowly moving, creeping, benthic juvenile (Plate 2C) that grows to be an adult. In all known cases, hatched veligers must feed and grow while planktonic (Perron 1981c). The life history thus includes an obligate minimum or precompetent planktonic period, during which growth and morphological and physiological development are rapid, and the larva is not able to metamorphose.

The maximum shell dimension of hatching *Conus* veligers varies by a factor of more than three, from 0.22 to 0.75 mm (coefficient of variation, CV $= 40$) in 18 species that Perron (1981a) studied in Hawaii. Size at metamorphic competence is more uniform, varying by a factor of 2 (shell length 1.1–2.1 mm; CV $= 21$) in 14 species reared through metamorphosis in the laboratory by Taylor (1975) and Perron (1981a: Fig. 1).

Perron and Kohn (1985) showed that the ordinary least squares (OLS) linear regression of the minimum or precompetent planktonic period (T_P) on egg diameter in the 13 species with both data known accounts for 92 per cent of the variance in the dependent variable. The authors used this regression as a model to estimate dispersal capability of *Conus* species across the spectrum of known egg sizes. We address the broader question of appropriate regression models in the methods section of Chapter 4. Here we provide fuller data on the sources of information given only in tabular form in that preliminary paper. We correct two identification errors ('*C. scabriusculus*' = *C. coffeae*; '*C. vidua*' = *C. araneosus*) and we provide comparable data for three additional species (*C. chaldaeus*, *C. geographus*, and *C. stramineus*).

The most striking external aspects of larval development and growth involve the velum, shell, and foot. The velum, which is the larval gastropod's swimming and feeding organ, undergoes a characteristic pattern of growth and development between hatching and metamorphosis. This pattern varies from species to species as a function of hatching size or egg size.

In species with small hatching sizes, e.g. *C. lividus*, veliger larvae with simple, rounded velar lobes emerge from the egg capsules at hatching. During planktonic development these lobes slowly enlarge and eventually become divided or bifurcate (Plate 2B). Veligers with larger hatching sizes (e.g. *C. striatus*) have velar lobes that are already bifurcate at the time of hatching (Plate 2C). Finally, species such as *C. pennaceus* (Plate 2D) hatch nonplanktonic veliconchas that emerge from the egg capsule with fully formed velar lobes similar to those of metamorphically competent planktonic larvae (Perron 1981*c*). The ontogeny of the velum in species with small hatching sizes thus consists of a series of stages that resemble the hatchlings of species with progressively larger hatching sizes. These observations support the hypothesis that larger eggs result in hatchlings that have progressed farther along a single developmental trajectory that is common to all species of *Conus*.

The hatchlings of species with small eggs and small veligers grow slowly in the plankton and have long precompetent planktonic periods. The larger hatching veligers of species with large eggs grow more rapidly and attain competence to settle and metamorphose sooner (Perron 1981*a*). This observed direct relationship between hatching size and planktonic growth rate is attributed to the developmental state of the velum at hatching (Perron 1981*c*). Since the food gathering capacity of the velum generally depends on the length of the ciliated groove that runs around its perimeter (Strathmann and Leise 1979), species that hatch with larger or more elaborately shaped velar lobes are likely to be capable of more efficient feeding and rapid growth.

The hatching veliger larva has a small, thin shell secreted during embryonic development in the egg capsule. This is the first protoconch, or Protoconch I. It has begun to spiral but usually comprises less than one complete whorl. During planktonic growth, the shell becomes conispiral; the whorls formed after hatching and prior to metamorphosis constitute Protoconch II. The larger Protoconch I is, the fewer and larger are the subsequent whorls of Protoconch II (Perron 1981*c*). As the shell enlarges, so does the foot, and the increasing weight of these two structures likely makes swimming with the velar cilia increasingly difficult. However, as Perron (1980) noted, major increases in shell weight are not apparent until immediately after metamorphosis when the shell begins to thicken and calcify.

The patterns summarized in this section suggest the hypothesis that the larger the egg, the farther the larger hatchling it develops into is advanced along a developmental trajectory that is common to all *Conus* species.

METAMORPHOSIS

The competent larvae of *Conus* settle and metamorphose preferentially on substrates with an organic film, but specific chemical cues do not seem necessary. At metamorphosis the velum is resorbed and the animal crawls on the enlarged, plantigrade foot (Perron 1980, 1981*c*) (Plate 2C).

In species with very large eggs (> 480 μm), hatching typically occurs at the veliconcha stage after a longer period of encapsulation than the species hatching as veligers. Hatchling veliconchas may still retain the velar lobes, but they also have a well-developed, functional crawling foot and a calcified shell. These two heavy structures dictate a primarily benthic habit, although some veliconchas (e.g. *C. pennaceus*) are capable of brief swimming. Of the species that hatch as veliconchas, only *C. pennaceus* has been studied in detail (Kohn 1961*a*; Perron 1981*c*). Hatchlings can metamorphose without food and thus may be considered lecithotrophic. They do eat and evidently utilize unicellular algae if available. Fed larvae prevented from metamorphosing survived for up to 39 days, 10 days longer than starved ones. At metamorphosis, veliconcha hatchlings are similar in shell dimensions to larvae that hatched as veligers and attained metamorphic competence while planktonic. They differ in shell form, however; the latter have more whorls (2.5–4.5 vs. 1.5) as well as smaller examples of Protoconch I (Plate 2D; Perron 1981*c*: Fig. 7).

In at least some *Conus* species, Protoconch I can be distinguished from Protoconch II at metamorphosis, because the former is calcified and turns opaque white while the latter remains translucent and yellowish or colourless. However, the boundary usually cannot be detected in the more heavily calcified shells of older individuals, because there is no clear external demarcation line.

Although shell length at metamorphosis among species of *Conus* ranges from 1.1 to 2.2 mm, this life history parameter is correlated with neither egg size nor hatching size (Perron 1981*a*). Hermans (1979) has documented a similar lack of correlation for polychaete annelids. This evidence suggests that the selective pressures that determine egg size and hatching size of pelagic larvae are not the same as those controlling size at settlement into the juvenile habitat and metamorphosis to the juvenile form. Evolution of egg size is likely to be influenced by such factors as pre- and posthatching mortality rates (Spight 1975, 1976), and the amount of energy adults are able to allocate to reproduction overall (Menge 1975; Todd 1979) and per offspring. The most appropriate size for the habitat switch from planktonic to benthic associated with metamorphosis is, however, more likely to be determined by the nature of the predators encountered in the juvenile habitat and by the food requirements of newly metamorphosed juveniles (Perron 1981*a*; Werner 1988).

From analysis of the literature on larval ecology of a broad range of prosobranch gastropods, Shuto (1974) proposed a quantitative relationship between the ratio of the diameter (in mm) of the larval shell or Protoconch II (D) at settlement and metamorphosis to the number of protoconch whorls (V) and the planktotrophy–lecithotrophy dichotomy, as follows (see also Jablonski and Lutz 1983):

Lecithotrophy $D/V > 1$ or $1 > D/V > 0.3$ and $2.25 > V$
Either $1 > D/V > 0.3$ and $3 > V > 2.25$
Planktotrophy $0.3 > D/V$ and $V > 3$

In Chapter 4, we examine the fit of *Conus* to Shuto's model, using both IP species and data from species studied elsewhere. We count the number of protoconch whorls from the origin of the suture, the practice of Shuto and of most subsequent workers (e.g. Burch 1980; Jablonski and Lutz 1980; Tursch and Germain 1985; Ponder and Clark 1989; Warén *et al.* 1989; Ramón 1990). Two other conventions that do not count the earliest part of Protoconch I are sometimes employed (Verduin 1982; Robertson 1985). These give lower counts by about ½ and ¼ whorl, respectively. We have used the method of counting from the origin of the suture because it was used by Shuto (1974) and his analysis comprises by far the largest and most general data set of any study; it is no less objective than the alternative conventions; and it is used in the majority of published reports.

Shell material added to the protoconch following metamorphosis constitutes the teleoconch; its continued growth forms the adult shell. In some prosobranchs, the boundary between Protoconch II and teleoconch is marked by a definite growth line, change of sculptural pattern, or both. In other taxa, no change in shell morphology is detectable at this boundary (Lima and Lutz 1990). In *Conus* there is no sharp demarcation, but rather the transition appears gradual over about 45° of shell growth, usually as microsculptural changes.

Because of the accretional and conispiral modes of shell growth, the protoconch and early teleoconch are retained at the apex of the shell as the gastropod grows. They provide a permanent record of early developmental history features throughout the animal's life. However, as the oldest part of the shell, the apex is exposed longest to environmental effects, and it is almost always eroded in *Conus* species from shallow, high energy habitats. The shells of juveniles from such habitats and of adults from deeper, more benign environments, provide the best preserved records of larval and early postlarval characters.

3 COLLECTIONS AND OBSERVATIONS

This chapter presents the basic data that are the sources of, and are essential to, the comparative analyses, syntheses and tests of the hypotheses in Chapters 4 and 5. They consist of descriptive accounts of the reproductive biology of each of the 62 IP species of *Conus* for which we have information and, in Appendix Table 1, a summary of the quantitative aspects in matrix form. This is to facilitate comparisons within and between species. As Underwood (1991) noted in another marine ecological context, publication of such large-scale surveys is important, because this provides the means to test hypotheses.

METHODS

The material reported on here was collected by the authors and by the late James E. Norton between 1956 and 1986, except where otherwise noted. Specimen numbers identify mothers; those collected by the first author of this text bear no prefix, those collected by the second author are prefixed FEP, and those collected by Norton are prefixed JEN. Other collectors are indicated in the text. Specimens in the collection of the Australian Museum are indicated by their catalogue numbers, prefixed 'C'.

All measurements were made on preserved material. Dimensions of egg capsules given include height of the capsule, excluding the stalk, and width, the maximum dimension perpendicular to the height (Kohn 1961a,b). The number of eggs in a capsule was determined by direct count and by estimation from counts in a Sedgwick–Rafter cell. Counts were usually made from several egg capsules in an egg mass. Egg diameters were measured with a dissecting microscope fitted with an ocular micrometer, usually at a magnification of $40\times$ ($80\times$ for the smallest eggs). Spacing of the micrometer grid was 24 μm at $40\times$, 12 μm at $80\times$. Because some estimation was thus required, ten eggs from each egg capsule were usually measured. Means and standard deviations are given in the Appendix, as are frequency distribution of coefficients of variation both within broods (Appendix Fig. 1A) and among individuals of each species (Appendix Fig. 1B).

Larval rearing studies were carried out at the Pacific Biomedical Research Center (PBRC) in Hawaii and at the Micronesian Mariculture Demonstration Center (MMDC) in Palau, Caroline Islands (Perron 1980, 1981a,b). Egg masses were maintained and monitored in laboratory sea-water tables until hatching appeared imminent. The egg capsules were then placed in 2-litre glass beakers filled with sea-water filtered at 5 μm. When hatching was observed, healthy larvae were transferred by pipette into 55-litre fibreglass culture tanks filled with filtered sea-water, except the larvae of *C. abbreviatus* and *C. lividus*, which grew best when reared in 2-litre glass beakers.

All larvae were fed a mixed phytoplankton culture made up of equal volumes of *Isochrysis galbana* and *Phaeodactylum tricornutum*. Each day, approximately 10 litres of water were siphoned off the bottom of the tanks. Since healthy larvae tended to remain swimming in the water column, dead larvae and debris could be removed from the tanks without significant loss of healthy animals. After the siphoning procedure, 50 mlitre of the mixed phytoplankton culture were added to each tank along with 10 litre of filtered sea-water. No antibiotics were used. Samples of larvae removed from the culture vessels at regular intervals were sacrificed for growth measurements. The maximum shell dimension was determined with an ocular micrometer, and growth curves were constructed for each species.

Larvae reared in the fibreglass tanks metamorphosed spontaneously on the bottom and sides of the tanks. Larvae reared in glass beakers metamorphosed only when placed on rocks or pieces of coral. Consequently, larvae of the two species mentioned above were tested daily for the ability to metamorphose as soon as signs of incipient metamorphic competence were detected. Such signs include enlargement and increased mobility of the foot and a tendency to swim close to the sides or bottom of the container.

The precompetent planktonic period (T_P) of each of the 11 species reared in the laboratory was defined as the minimum time required for larvae to develop from hatching to metamorphosis. These data, along with measurements of egg diameter for each species, were used to derive equation (3.1) below. This equation was then used as a model to estimate the planktonic periods of species for which no larval rearing data were available (Perron and Kohn 1985).

The length of the precompetent planktonic period was estimated from egg diameter by the OLS regression (Eqn 3.1), derived from 11 *Conus* species reared through settlement and metamorphosis (from Perron and Kohn 1985, with minor corrections):

$$Y = 40.8 - 0.087X \quad (r^2 = 0.92) \quad (3.1)$$

where Y = precompetent period in days and X = egg diameter in μm. In a few cases, egg diameter was estimated from hatching size using Eqn 3.2, an OLS regression derived from 19 *Conus* species reared through hatching in Hawaii (corrected from Perron 1981a):

$$X = \frac{Y + 167.3}{2.67} \quad (r^2 = 0.96) \quad (3.2)$$

where X = egg diameter and Y = maximum shell dimension at hatching, both in μm.

For each species, we briefly indicate previously published records, and we then list new records individually. The latter include records of egg diameter and estimated length of precompetent planktonic periods that we listed in a preliminary paper without further data (Perron and Kohn 1985: Table 1). Finally, for species with numerous records we summarize as many of the following attributes as are known for each species: egg size and number, egg capsule and capsule mass features, oviposition site, course of embryonic development, and observed or estimated planktonic period necessary prior to metamorphic competence. For species with adequate data, we examine the relationships of these attributes with body size of the mother, expressed as shell length. Fuller details of all records are given in tabular form in the Appendix.

SPECIES ACCOUNTS

Conus abbreviatus Reeve

Prior records

Ostergaard (1950) briefly described and illustrated the egg capsules of *C. abbreviatus* but gave few details. Kohn (1961a) described and illustrated the stages of embryonic development, and Perron (1981a,b; 1982) reared larvae through metamorphosis and analysed energetics of reproduction in *C. abbreviatus*.

Egg capsules up to 10 × 8 mm, containing about 1300 eggs, 170 μm in diameter, are deposited in a Type I egg mass on the undersides of objects on reef platforms. The egg mass studied in most detail contained about 44 000 eggs. Planktotrophic veliger larvae, 270–300 μm in maximum shell dimension, hatch 14–15 days later. Settlement and metamorphosis occurs after a 32-day precompetent period during which the shell grows to 1.1 mm.

Conus achatinus Gmelin

New records

1. Lee Pt., near Darwin, NT, Australia (12° 20' S, 130° 54' E); 20 August 1970. Coll. J. B. Cameron. AMS No. C.158483. Four capsules were collected from an egg mass with two adult females and a male. All contained uncleaved eggs 805–902 μm (mean = 837 μm) in diameter.

2,3. Waigait, near Darwin, NT, Australia (11° 28' S, 130° 50' S); 7 September 1970. Coll. O. J. Cameron. AMS No. C.158483. Two egg masses were collected on a sheltered rocky reef.

One consisted of 21 egg capsules with an adult female. They contained well-developed veliconchas with shells 1.5 mm long and of two whorls. The other egg mass, found with an adult male, consisted of 27 capsules containing earlier veliconchas with shells about 1 mm long.

Conus achatinus has the largest eggs (average in all samples, 824 µm) of any known Indo-Pacific species of *Conus*, and the number of eggs in the Type I egg mass, estimated at 714–775, is among the smallest.

Conus ammiralis Linnaeus

New records

1. Lagoon reef flat near airbase, Kwajalein Atoll, Marshall Islands; 1 August 1983. No. FEP-277. Coll. L. Boucher and S. Johnson. An adult female found crawling on sand at night was transported alive to the MMDC, where it was kept in a flowing sea-water system and fed specimens of *Cypraea annulus*. This is the first confirmation of the molluscan diet of *C. ammiralis*. A partially completed egg mass was discovered on 12 August 1983, and development was monitored until hatching occurred on 24 August 1983. The egg mass consisted of individual capsules in a three-dimensional cluster with many bridges between them.

2. Lagoon reef, Uliga I., Majuro Atoll, Marshall Islands; 8 September 1956. No. 3898. Oviposition was interrupted after two egg capsules had been deposited on the underside of a coral boulder.

3. Wading I., W of Viti Levu, Fiji, February 1962. Coll. A. Jennings. All but one of the capsules in the egg mass were affixed to the other capsules.

The compact Type II egg masses of *C. ammiralis* are deposited on the undersides of coral rocks in lagoon environments. Egg capsules are 10–15 × 7–10 mm; most are attached to others rather than directly to the substratum. Each capsule contains about 270–520 eggs, 300–350 µm in diameter. Development has not been followed, but hatching veligers had shells 660 µm in maximum dimension. The minimum precompetent period is estimated from egg diameter at 11–14 days.

Conus araneosus Solander

Previous record

Natarajan (1957) reported mating habits and described in detail and illustrated egg capsules and developmental stages of *C. araneosus* from Vedalai, near Mandapam Camp, India.

New record

S of Popototan I., Calamian Group, Palawan, the Philippines; 12 December 1963. No. JEN-14. A female with egg mass was collected on a sand and mud bottom at a depth of 4 m.

As in other molluscivorous species the *C. araneosus* egg mass is of Type II; most egg capsules of *C. araneosus* are affixed to previously laid capsules, in this case in two or three layers. Natarajan (1957) did not observe natural hatching, but veligers 1100 µm in shell diameter released from egg capsules swam and crawled for up to 36 h. The egg capsules from the Philippines contained fewer (average of 43 vs. 164) and somewhat larger (517 vs. 492 µm) eggs than those from India, suggesting the absence of a planktonic larval phase.

Conus arenatus Hwass in Bruguière

Prior record

Kohn (1961b) described three egg masses of *C. arenatus* from the western Indian Ocean. Since all contained late developmental stages, egg size of this species remains unknown. The egg mass is of Type I, but a few capsules are attached to previously deposited capsules. The egg capsules measured 10–14 × 6–11 mm; each contained 1000–5300 embryos. The egg mass was estimated to contain about 65 000 eggs.

Conus aristophanes Sowerby

This species is very closely related to, and often synonymized with, *Conus coronatus* Gmelin (e.g. Kohn and Nybakken 1975). Populations of the two forms in Fiji are quite distinct, however;

Cernohorsky (1964) indicates seven distinguishing characters and considers them separate species.

New record

Talira, Tai Levu coast of Viti Levu, Fiji; 9 December 1986. No. 11459. Coll. B. Holthuis. A female 25 × 16 mm and two egg capsules 6.5 × 6.5 mm were collected at a depth of 1−2.5 m. The capsules contained about 1300 uncleaved ova 186 μm in diameter. A precompetent planktonic period of 25 days is estimated from egg diameter.

Conus aulicus Linnaeus

New record

Pulo Penju, NE of Pulo Simalur, Sumatra, Indonesia; 22 November 1963. No. 6147. A specimen 96 × 40 mm was collected from a large crevice in a coral rock with a mass of 79 egg capsules 19−27 × 15−19 mm, each containing about 1600−2200 well-formed veligers. As in other molluscivorous *Conus* species, the egg mass is of Type II. Further development could not be followed, as all veligers died 3 days later. A precompetent planktonic period of 12 days is estimated from the diameter of uncleaved eggs present in some of the capsules.

Conus balteatus Sowerby

New record

Pulo Bai, Batu Group, Sumatra, Indonesia; 25 November 1963. No. 6319. A female 37 × 23 mm was observed ovipositing on the underside of a large coral head in 1 m of water on the reef platform. Capsules in the Type I egg mass 11 × 10 mm contained 420−450 eggs of mean diameter 311 μm. Development could not be followed for more than 4 days, at which time the embryos had attained the trochophore-equivalent stage. The precompetent planktonic period is estimated from egg diameter at 14 days.

Conus bandanus Hwass in Bruguière

Prior record

Perron (1981*a*,*b*,*c*) reported on the larval life history of this species (as *C. marmoreus*) from Hawaii. Eggs 344 μm in diameter hatched as planktotrophic veligers 755 μm in maximum shell dimension. Settlement and metamorphosis occurred in the laboratory after a minimum pelagic period of 10 days, during which shell length reached 1.5 mm.

New record

Ngadarak Reef, entrance to Malakal Harbor, Palau Islands; 22 October 1983. No. FEP-280. A female 59 × 34 mm was collected under a boulder to which the egg mass was attached. The Type II egg mass was identical in form to that of the closely related species *C. marmoreus* (q.v.). The egg capsules were open and hatching was in progress at the time of collection. Hatching veligers were 630 μm in maximum shell dimension. Egg diameter is estimated from hatching size at 300 μm, and the minimum planktonic period is estimated to be 15 days.

Conus biliosus (Röding)

Prior record

Zehra and Perveen (1991) observed spawning in nature and in the laboratory between April and June in Pakistan. Individual females spawned twice on average during the season. About 24 egg capsules 11 × 8.5 mm are deposited in rows of 4−6 in a Type I egg mass. Table 3.1 summarizes the observations of Zehra and Perveen (1991) on embryonic development, which lasted about 11 days and resulted in hatching of veligers about 250 μm in maximum shell dimension. Larvae were negatively geotactic and positively phototactic immediately after hatching. They survived up to 62 h in the laboratory. Minimum planktonic duration is estimated to be 26 days.

Table 3.1 Summary of new records of development to hatching in *Conus* species. Numbers in body of table indicate day (or hour in parentheses) following oviposition (roman type) or collection (italic type). Data on localities and egg mass characteristics are in Appendix Table 1

Species	Number	Uncleaved egg	2-cell	4-cell	8-cell	Blastula	Gastrula	Early trochophore-like	Trochophore-equivalent	Early veliger	Well-developed veliger	Hatching
biliosus	*						4–5	5	5.5	6–7	8	10–11.5
canonicus	6168	(6)	(7)		(12)	4					7 (171)	
canonicus	6169		(5½)		(5½)						3 (80)	
canonicus	6397									*(10–60)*		
canonicus	6912			(15–25)			3 (70)	4	6	7–8		4
capitaneus	6032	*(11)*			—3 (38)—			3		5	7	8
capitaneus	6033				—3 (38)—						3½	7
catus	6069–6070		(4½)				2 (52)	4 (90)	6	7	7	
coronatus	*		(6)	(7)	(12)	4	4–5	5	5.5	6–7	7	9.5–10
ebraeus	7164	*(10)*		(8)	(33)				1–2 (26–55)	3 (82)	9	13
eburneus	8485–8486									2–4	6	10
episcopus	6367							*(6)*				
frigidus	6864		—(10)—				3 (74)		4 (103)	6 (135)	9–10	
frigidus	7150		—(10½)—	—(27)	2 (55)							
glans	FEP-217											14
lividus	6151								(4½)	1½ (38)	3 (86)	
magus	6238	(13)				2½ (62)			6½ (155)		12 (285)	15
magus	6240	(13)		—2½ (62)—			4½ (107)					
miliaris	6031	(2½)					*1–2 (25–52)*		*3½ (86)*	*6–7*	*8 (196)*	*9*
striatus	—		(4)			3 (68)			—6 (138)—	9–12	14	
textile	6126	(6½)					3					
textile	6129						*(6½–15)*	2 (60)	4 (100)	8 (198)	10	

* Data from Zehra and Perveen (1991).

Conus canonicus Hwass in Bruguière

Prior record

Kohn (1961b) described larval development in two egg masses from the Seychelles referred to as *C. textile* 'with some hesitation', as the adults differed 'from typical *C. textile* by having a narrower and thicker shell with a pink aperture'. It is now clear that these were specimens of *C. canonicus*.

New records

1,2. Pulo Penju, NE of Pulo Simalur, Sumatra, Indonesia; 22 November 1963. Nos. 6168, 6169. Two adults, one ovipositing on the underside of a coral rock in 0.3 m of water, and one on sand under a coral rock in 0.7 m of water with a mass of 35 egg capsules. Two of the latter were fixed to the substratum, with the others fixed to these two and to each other in a Type II egg mass typical of species of the *C. textile* group. When examined 4 days after collection, the latter capsules contained well-developed veligers, with eyes and foot with operculum present (Fig. 3.1D). Each velar lobe was pigmented with a dark peripheral line and a row of about nine green pigment spots proximal to it. All of these veligers died without hatching by the ninth day after collection. On the seventh day following collection, the egg capsules collected during oviposition contained very early veligers (Fig. 3.1B); all died as advanced veligers without hatching two days later.

3. Pulo Bai, Batu Group, Sumatra, Indonesia; 27 November 1963. No. 6397. A female with a mass of 54 egg capsules was collected on sand under a live *Porites lutea* head. The capsules contained trochophores and early veligers (Fig. 3.1C). The latter had small velar lobes with few pigment spots, very small eyespots, and a small foot with operculum. On the fourth day following collection, some capsules still contained veligers about 343 μm in maximum dimension, while others had begun hatching veligers that now measured 417 μm.

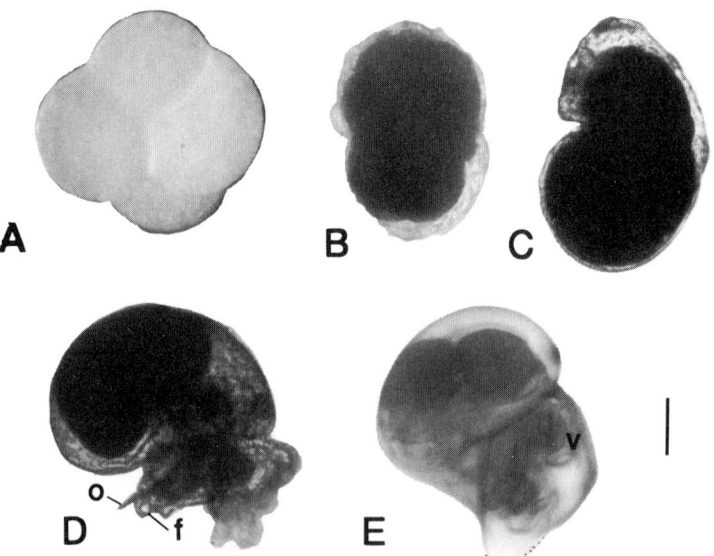

Fig. 3.1 Developmental stages of *Conus canonicus* Hwass in Bruguière. Nos. 6912 (A), 6168 (B), 6397 (C,E), 6169 (D); Indonesia. (A) 4-cell stage, 25 h after oviposition. (B) Very early veliger stage, 7 days after oviposition. (C) Early veliger. (D) Well-developed veliger. (E) Hatched veliger. Scale bar = 100 μm. f, Foot; o, operculum; v, retracted velar lobe with dark peripheral band.

4. Sanding I., Indonesia; 5 December 1963. No. 6912. A female 42 × 20 mm was collected with a mass of 37 egg capsules. They contained mainly 4-cell stage embryos averaging 369 × 334 μm (Fig. 3.1A). By the following day the third and fourth cleavages occurred. Embryos were in gastrula stage on the third and fourth days following collection. Late trochophore-equivalent stage occurred on the sixth day, and early veliger stage, with small velar lobes and eyes, very short tentacles, and foot with operculum, on the seventh day. By the tenth day following collection, all embryos were dead in the veliger stage; no hatching occurred.

Table 3.1 summarizes the observations on embryonic development in the four *C. canonicus* egg masses from Indonesia. None of these survived to hatching, but Kohn (1961b) noted the prehatching period to be 14–15 days.

Despite the rather small sample of *C. canonicus* egg masses, the data were sufficient to indicate positive relationships of adult body size with egg capsule size and with number of eggs per capsule (Fig. 3.2A, Table 3.2). Egg diameter is not significantly correlated with adult body size ($r_s = 0.40$; $P << 0.05$) (Fig. 3.2A, Table 3.2).

The egg mass of *C. canonicus* is of Type II. Mean egg diameters in the four egg masses from Indonesia were 258–278 μm, indicating an estimated precompetent planktonic period of 17–18 days. The number of eggs per capsule ranged from 450 to 1400, and the estimated number of eggs per egg mass, from 17 000 to 76 000.

Conus capitaneus Linnaeus

Prior record

Risbec (1932) reported two Type I egg masses of *Conus capitaneus* from New Caledonia. He described and illustrated the capsules and the arrangement of the eggs within, but he gave no quantitative information other than that one of the egg masses comprised 64 capsules at the time oviposition was interrupted.

New records

1,2. Goh Sindarar Nua, Thailand; 7 November 1963. Nos. 6032, 6033. Two adults were collected on sand under conglomerate boulders, with egg capsules attached to the undersides of the rocks. When examined 11 h after collection, the capsules of No. 6032 contained spherical eggs with the first two cleavage furrows visible. When examined 38 h after collection, embryos of both egg masses ranged from 2- and 4-cell to gastrula stage. Three days after collection, No. 6032 embryos were at early trocophore-equivalent stage; they attained early veliger stage 5 days after collection. Well-developed veligers with eyes and otoliths present were observed 7 days after collection. These hatched, perhaps prematurely, on the eighth day after collection (Table 3.1). The shells of the

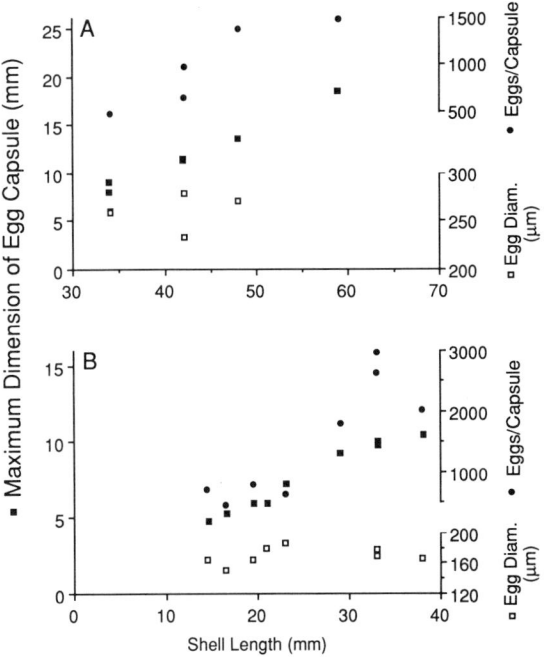

Fig. 3.2 Relationships of egg diameter, egg capsule size, and number of eggs per capsule with body size of mother (as shell length). (A) *Conus canonicus* Hwass in Bruguière. (B) *Conus coronatus* Gmelin. For correlation coefficients, see Table 3.2.

Table 3.2 Correlations between adult size and spawn in Indo-Pacific *Conus*. Spearman rank correlation coefficients (r_s) are given for species with $N \geq 5$ egg masses. See Appendix Table 1 for complete data. *, $P < 0.05$; **, $P < 0.01$; other r_s values not significant at $P_{0.05}$

	Adult size vs.						
	Capsule size		Eggs per capsule		Egg diameter		Size range
Species	N	r_s	N	r_s	N	r_s	of ♀♀ (mm)
biliosus[1]	6	0.93*	—		—		25–33
canonicus	6	0.99**	6	0.97*	5	0.40	34–59
catus	5	0.53	5	0.72	5	−0.33	30–40
consors	6	0.56	5	0.68	5	0.09	40–57
coronatus	9	0.99**	8	0.80**	9	0.46	14–38
coronatus[1]	6	0.96*	—		—		16–21
leopardus	5	0.94*	5	0.40	5	0.17	80–140
lividus	14	0.85**	7	0.96**	12	0.42	24–66
magus	11	0.39	10	0.53	10	0.16	35–52
marmoreus	6	0.94**	5	1.00**	6	−0.90*	54–95
pennaceus	7	0.93**	5	0.80	6	−0.27	33–60
rattus	5	0.72	4	0.15	5	0.05	27–33
textile	11	0.88**	8	0.85**	8	−0.05	54–80

[1] Data from Zehra and Perveen (1991).

hatched veligers measured 290 μm in maximum dimension; expected shell size at hatching from Eqn 3.2 is 228 μm. Eggs of No. 6033 developed even more rapidly, attaining rather well-developed veliger stage 3½ days after collection. They hatched on the seventh day after collection (Table 3.1), but all had died 2 days later. If it is assumed that these egg masses were collected on the day following oviposition, the prehatching period of 8–9 days is shorter than any previously reported for *Conus* (Perron 1981a).

3. Ngadarak Reef at entrance to Malakal Harbor, Palau; 24 May 1982. No. FEP-30. An adult female was observed ovipositing on the underside of a dead coral slab, behind the reef crest in 0.5 m of water. The eggs were not cultured.

Conus capitaneus egg masses are of Type I, with the capsules individually affixed to the substratum in rows of up to 7 capsules. The bases of most capsules in each row are confluent. About 3000–5000 eggs are in each capsule.

The precompetent planktonic period is estimated from the egg diameter to last 27–29 days.

Conus catus Hwass in Bruguière

Prior records

Ostergaard (1950) described and illustrated the Type I egg mass of *C. catus* from Hawaii and followed development from uncleaved egg through hatching 15–16 days later. Kohn (1961a) briefly described properties of three *C. catus* egg masses also from Hawaii.

New records

1. Koko Head, Oahu, Hawaii, 22 August 1956. No. 2298. An adult was observed ovipositing.

When examined, most embryos were in the 4-cell stage. This observation was inadvertently omitted from Kohn (1961a).

2. Goh Huyong, Similan I., Thailand; 8 November 1963. Nos. 6069–6070. Two adult females were collected, one adhering to the underside of a coral rock with a mass of 35 egg capsules, the other mostly buried in sand under the same rock. When examined 4½ h after collection, eggs were in the 4-cell stage (Table 3.1). On the sixth day following collection, a cup-shaped shell had formed and the ciliated foot rudiment was present. Early veliger stage was attained the following day, but by the next day all had died.

3. Talaga Cove, Bataan, the Philippines; 28 February 1964. No. JEN-56. A female and egg mass were collected from the underside of a rock at a depth of 7 m. Most embryos were at the 4-cell stage, but some uncleaved eggs were present.

4. Reef flat behind Ngadarak Reef, Palau; 10 June 1982. No. FEP-53. A female was observed ovipositing on the underside of a dead coral slab.

5. Alofi, Niue; 11 September 1986. No. 11455. Coll. B. Holthuis. A female with an egg mass of 25 capsules was collected under a rock in a tide pool on the reef flat. Preserved shortly after collection, the capsules contained uncleaved eggs, 2- and 4-cell stages.

Conus catus eggs are 200–240 μm in diameter; about 800–1600 are in each capsule. The egg mass is of Type I. Egg capsule size and number of eggs per capsule were positively but not significantly correlated with adult shell length (Table 3.2). The duration of the precompetent planktonic stage is estimated from egg diameters to be 20–23 days.

Conus chaldaeus (Röding)

New record

Namui Reef, Niue; 15 September 1986. No. 11456. Coll. B. Holthuis. A female with a few egg capsules was collected from the underside of a rock in a tidepool on the reef flat. The capsules are smooth except for corrugated distal margins, as in the closely related species *C. ebraeus* (q.v.). When preserved, the capsules contained uncleaved ova with a mean diameter of 175 μm. The estimated duration of the precompetent planktonic period is 25 days.

Conus cinereus Hwass in Bruguière

New records

1. 0.3 km SW of San Fernando, La Union, the Philippines; 25 April 1964. No. JEN-44. A mass of 56 egg capsules was collected on sand at a depth of 7 m with an adult female. Most embryos were at the 2-cell stage when preserved.

2. Maniuayan I., Marinduque, the Philippines; 8 April 1964. No. JEN-18. A very large egg mass attached to a waterlogged stick was collected; an adult *C. cinereus* was collected from an adjacent rock. Almost all of the capsules were empty, but a few contained advanced veliconchas 660–760 μm in shell length.

3. 1 km N of Taytayen Pt., Batangas, the Philippines; 20 January 1964. No. JEN-57. A mass of 89 capsules attached to waterlogged sticks was collected at a depth of 9 m with an adult female. Some capsules contained uncleaved eggs that averaged 496 μm in diameter; others contained 2- and 4-cell embryos.

The egg capsules of *C. cinereus* measure 9–12 × 7.5–12 mm; the egg mass is of Type I. Each contains 100–200 eggs nearly 0.5 mm in diameter. Although nearly 100 capsules may be deposited in an egg mass, the two egg masses contained fewer than 10 000 eggs.

The egg size and developmental mode indicates that *C. cinereus* lacks a planktonic stage in its life history.

Conus coffeae Gmelin

New record

Western barrier reef 1 km S of Toagel Mlungui Channel, Palau, Caroline Islands; 6 November 1982. No. FEP-135. A female was observed ovipositing deep in *Acropora* rubble.

Formerly known as *Conus scabriusculus*, *C. coffeae* (see Coomans and DeVisser 1987) shares the habitat with the similar *C. glans*. Although the two species have virtually identical egg capsules, the ova of *C. coffeae* are much smaller than those of *C. glans*, and more are present per capsule. Duration of the precompetent planktonic stage is estimated from the egg diameter to be 21 days.

Conus consors Sowerby

New records

1. Uliga I., Majuro Atoll, Marshall Islands; 10 September 1956. No. 3946. An adult female was observed ovipositing on the underside of a coral boulder in 1 m of water in the atoll lagoon.

2. Lagoon, Japtan I., Enewetak Atoll, Marshall Islands; 1 September 1956. No. 3495. A female was on sand under a rock in 1 m of water. A cluster of 18 egg capsules was attached in three rows to the underside of the rock. The capsules, which contained uncleaved ova, are distinguished by the triangular shape of the exit window (Fig. 3.3), in contrast to the elongate windows with parallel sides of other *C. consors* and of other species examined.

3. Lagoon, Ponape, Caroline Islands; 26 August 1956. No. 3348. A female was found partly buried in sand under a coral rock to the underside of which egg capsules were attached. All had hatched, but a few contained veliger larvae about 0.6 mm in maximum shell dimension.

4. 1.1 km NW of Portuguese Pt., Lingayen Gulf, Pangasinan, the Philippines; 20 April 1964. No. JEN-3. An adult female was collected with a mass of 35 freshly deposited egg capsules in 5 m of water.

5. W of Cabatitian I., Lingayen Gulf, Pangasinan, the Philippines; 21 April 1964. No. JEN-32. An adult female with egg capsules was collected on sand at a depth of 4 m.

6. Malakal Harbor, Palau, Caroline Islands; 27 May 1982. No. FEP-35. An adult female with egg capsules was collected in sand under a rock on a patch reef.

7. Sand flat behind Ngadarak Reef, Palau, Caroline Islands; 30 May 1982. No. FEP-36. An adult female was collected with an egg mass under a rock.

8. Malakal Harbor, Palau, Caroline Islands; 14 June 1983. No. FEP-62. A female with egg capsules was collected in sand under a rock in 1.5 m of water on the fringing reef near the MMDC.

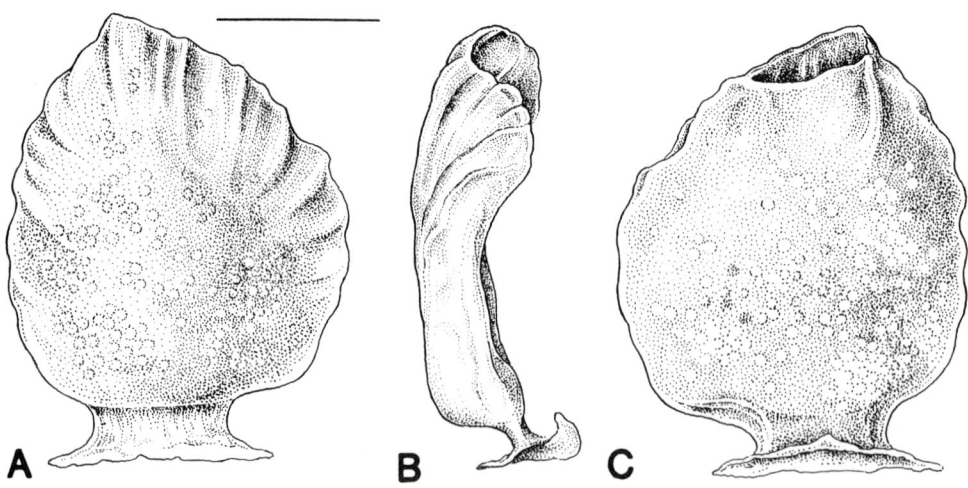

Fig. 3.3 Three views of an egg capsule of *Conus consors* Sowerby. No. 3495; Marshall Is. In (A) and (C), outlines of individual eggs are visible through the capsule wall. Scale bar = 5 mm.

The egg capsules of *C. consors* are 10–15 × 7.5–13 mm and are typically affixed in rows to the underside of a rock or coral head. The capsules from new records Nos. 1,2,4,5, and 8 contained about 150–300 uncleaved ova 390–440 μm in diameter; the other egg masses contained later developmental stages. A few uncleaved ova were usually present even in capsules containing hatching stage larvae.

In laboratory culture, the larvae of *C. consors* grew rapidly when fed on phytoplankton and attained metamorphic competence 7 days after hatching. Unfed larvae grew only slightly and never underwent metamorphosis. Although competent larvae of *C. consors* were frequently observed to metamorphose spontaneously on the fibreglass sides of the rearing vessel, much higher rates of metamorphosis were obtained by placing larvae directly on rocks which had been submerged in flowing sea-water for several days. Rates of successful metamorphosis obtained in this way ranged between 26 per cent and 84 per cent of larvae tested.

The Type I, *C. consors* egg masses from the Philippines and Marshall Islands contained somewhat larger eggs (410–440 μm vs. 390 μm) than those from Palau. Egg capsule size and number of eggs per capsule are positively, but not significantly, correlated with adult size (Table 3.2). The duration of the precompetent planktonic period of the former is estimated from egg diameter at 3–5 days.

Conus coronatus Gmelin

Prior records

Kohn (1961*b*) reported on six egg masses of *C. coronatus* from the Maldives, Seychelles, and Cosmoledo Island in the Indian Ocean. Mean egg diameters ranged from 150 to 180 μm. It was not possible to follow embryonic and larval development, but veliger shells from partially hatched capsules were 240–250 μm in maximum shell dimension.

Barkati and Ahmed (1985) and Zehra and Perveen (1991) reported that in Pakistan, *C. coronatus* spawns from April to June; the latter authors state that individual females spawned two or three times during the breeding season. Zygotes developed to the blastula stage in 4 days and well-developed veliger stage in 7 days. Hatching occurred at 9–10 days, at 32 °C. Newly hatched veligers were negatively geotactic and positively phototactic. They survived only 27 h in the laboratory; the estimated minimum planktonic duration is 27 days.

Huish (1978) briefly noted egg size and number of eggs per capsule in a *C. coronatus* egg mass from One Tree Reef, Queensland, Australia. He also reported collecting egg capsules of *C. coronatus* at Minnie Water (30° S), New South Wales, Australia, in June, 1978.

New records

1,2. S shore of Mandapam, India; 23 February 1968. Nos. 7384, 7385. Two females were collected with egg masses under rocks at the jetty of the Central Marine Fisheries Research Institute. Both egg masses contained mainly uncleaved ova and early cleavage stages. It was not possible to follow the course of embryonic development.

3. Avarua, Rarotonga, Cook Islands; 9 October 1984. No. BRAR-68. Coll. G. Paulay. An adult female was observed ovipositing under a rock on the fringing reef.

The small egg capsules of *C. coronatus* (4–12 × 4–12 mm), containing 400–3000 eggs, are affixed to the undersides of rocks in a Type I egg mass. Egg capsule size and number of eggs per capsule are highly significantly correlated with adult size. Egg diameter is positively but not significantly correlated with adult size (Fig. 3.2B; Table 3.2). The egg diameters from the records listed above indicate a precompetent planktonic period of 24–28 days.

Conus dictator Melvill

Prior record

Thorson (1940*a*) described the egg capsules and development of this species (reported as *C. planiliratus* var. *acutangulus*) from the Persian Gulf.

Conus planiliratus (= *C. inscriptus* Reeve; Kohn 1978) is a different species from that illustrated by Thorson (1940a: Fig. 24K). Although an earlier name may exist for this species, we presently consider it to be *C. dictator* Melvill, described in 1898 from the Persian Gulf (Korn 1990). The large eggs (575 μm diameter) and development to a veliconcha stage exceeding 1 mm in shell length within the egg capsule indicate that this species lacks a planktonic larval stage.

Conus distans Hwass in Bruguière

New record

Malakal Reef Crest, Palau, Caroline Islands, 15 August 1984. No. FEP-397. An adult female deposited two egg capsules in the laboratory five days after collection. Both were sacrificed for egg counts.

Conus ebraeus Linnaeus

Prior records

Risbec (1932) first described the egg mass of *C. ebraeus*, from New Caledonia, but he did not follow development. Ostergaard (1950) and Perron (1981a) reared *C. ebraeus* through hatching in Hawaii. Hatched veligers measured 280 μm in maximum shell dimension. Kohn (1961b) reported the characteristics of the egg mass of this species from Cosmoledo Island in the Indian Ocean.

New records

1. Parry Island, Enewetak Atoll; 19 August 1956. No. 3133. An adult (of undetermined sex) with four egg capsules affixed to the antero-dorsal part of the shell (Fig. 3.4) was collected on the intertidal seaward reef platform. The capsules contained early veliger larvae measuring 155–185 μm in maximum shell dimension. Although the first author has studied this site intensively (Kohn and Leviten 1976; Leviten and Kohn 1980; Kohn 1987), *Conus* egg capsules are rarely observed there. The environment is physically harsh (Kohn 1987) and

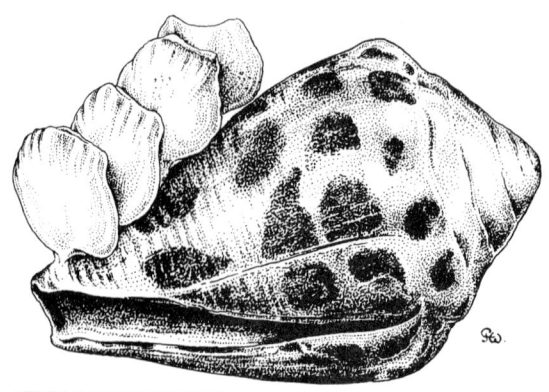

Fig. 3.4 Egg capsules of *Conus ebraeus* Linnaeus. No. 3133; Marshall Is. Capsules are attached to a *C. ebraeus* shell. Scale bar = 1 cm.

sheltered sites appropriate for oviposition are likely in short supply.

2. Sanding I., S of Sumatra, Indonesia; 7 December 1963. No. 7164. A female was observed ovipositing on the underside of a coral rock in 0.8 m of water. A capsule fixed 10 h later contained uncleaved ova and some with the first two cleavage furrows visible. A capsule examined 26 h after collection contained unexpectedly advanced embryos, in motile, trochophore-equivalent stage. All embryos died without further development.

The Type I egg mass of *C. ebraeus* contains capsules 6–10 × 5–10 mm in diameter (Fig. 3.5). The duration of the precompetent planktonic stage is estimated from egg diameter to be 25–27 days.

Conus eburneus Hwass in Bruguière

New records

1. Japtan I., Enewetak Atoll, Marshall Islands; 1 September 1956. No. 3488. A cluster of egg capsules was found affixed to the underside of a small coral rock in 1 m of water on the lagoon side of the island. An adult female was completely buried in sand under the same rock. One egg capsule (Fig. 3.6) contained embryos in the trochophore-equivalent stage measuring 185 ×

Plate 1 (A) *Conus nobilis*, living adult. (B) Copulation in *Conus tulipa*. The male is above, facing right; the female below, facing left. The penis is seen emerging from just under the edge of the male's shell. (C) Female *Conus tulipa* ovipositing on the wall of an aquarium. (B) and (C) Kwajalein Atoll, Marshall Islands. Photos by G. Pearson. (D) *Conus pennaceus*, female with egg mass on the underside of a coral rock. Eggs (500 μm in diameter) are visible through the exit window of some egg capsules. Oahu, Hawaii.

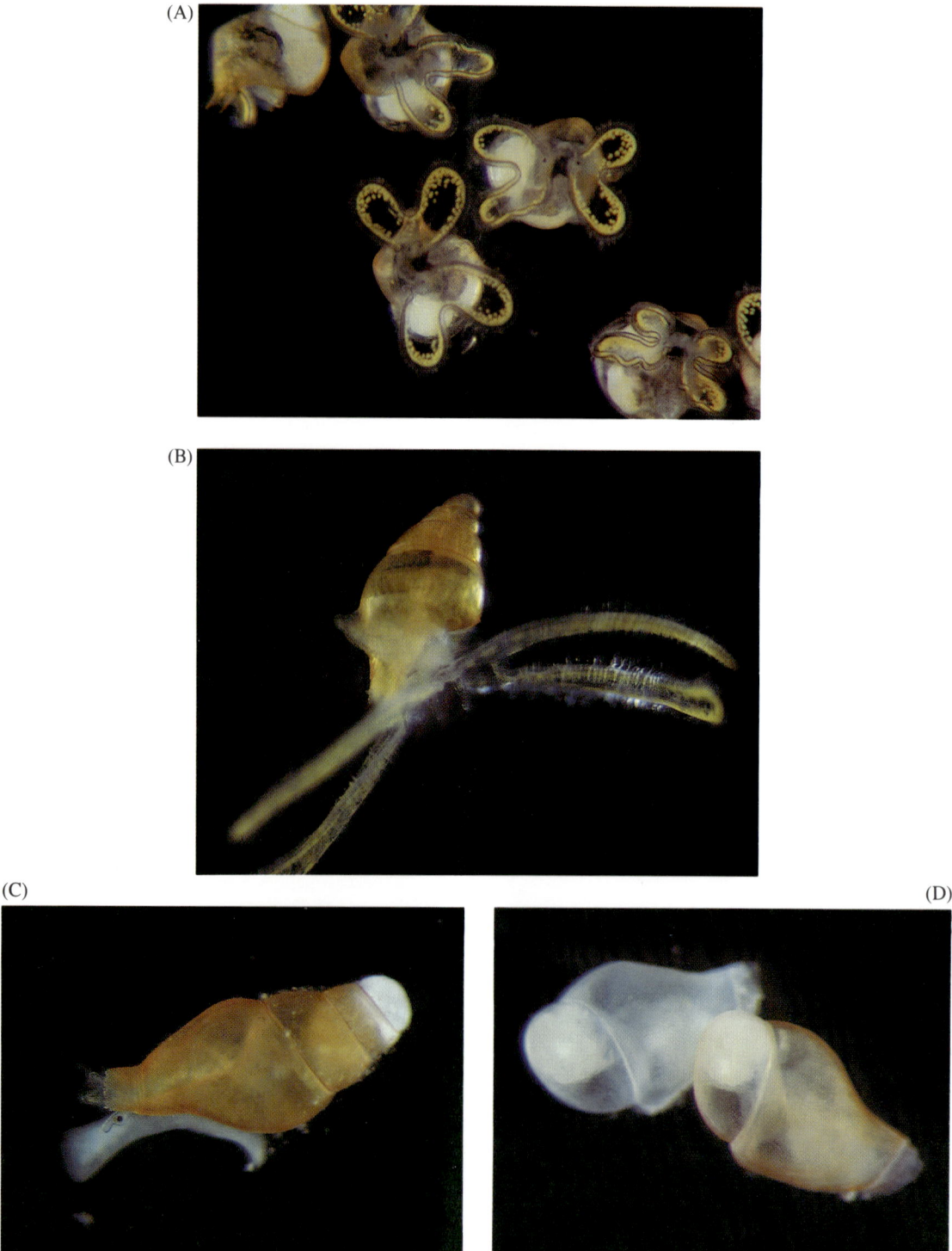

Plate 2 (A) *Conus marmoreus*. Recently hatched planktonic veligers. (B) *Conus lividus*. Older planktonic veliger, with long, bilobate velar lobes and multi-whorled shell. (C) *Conus striatus*. Metamorphosed juvenile from planktonic larva. (D) *Conus pennaceus*. Metamorphosed juvenile from non-planktonic veliconcha.

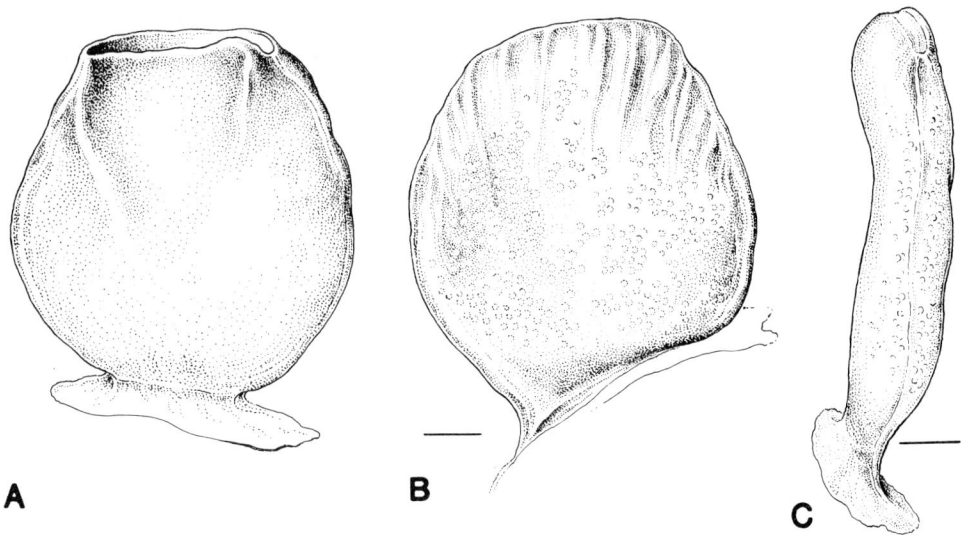

Fig. 3.5 Three views of an egg capsule of *Conus ebraeus* Linnaeus. No. 3133; Marshall Is. Scale bar = 1 mm.

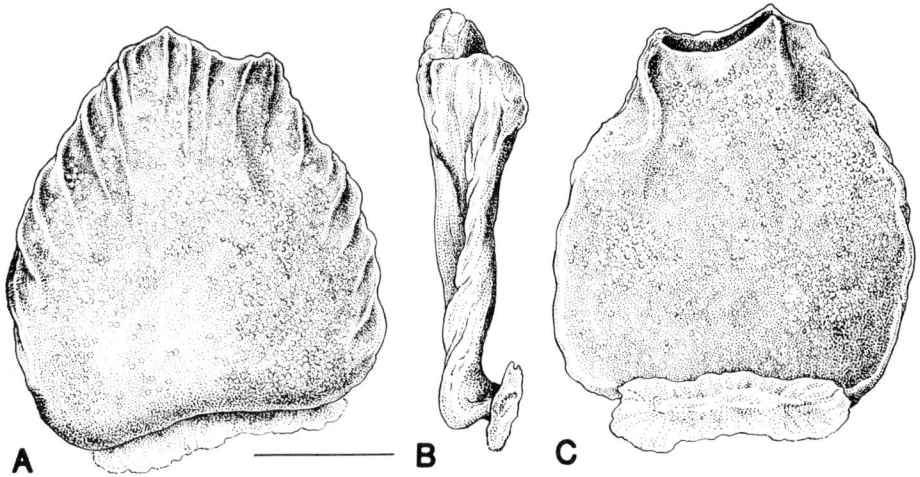

Fig. 3.6 Three views of an egg capsule of *Conus eburneus* Hwass in Bruguière. No. 3488; Marshall Is. Scale bar = 5 mm.

140 μm. Based on these measurements an egg diameter of about 150 μm is likely.

2. Heron I., Queensland, Australia; 18 February 1975. Nos. 8485, 8486. An egg mass was collected within 20 cm of two partly buried adults in 0.8 m of water on the inshore portion of the reef fringing the north shore of Heron I. The substratum was sand with abundant tubes of chaetopterid polychaetes. Most of the 105 capsules were attached to other capsules, forming a tight, three-dimensional

Type II mass as in many molluscivorous *Conus* species. The first capsules to be deposited, however, were attached to a broad strand of mucus to which sand particles adhered. The egg capsules of *C. eburneus* contain 1000–5000 eggs 150–210 μm in diameter. When first examined 8 h after collection, the embryos were at the 4-cell stage. It is thus likely that they were collected within 3 days of oviposition. Development was followed (Table 3.1), and veligers hatched on the 13th day after collection. Duration of the precompetent planktonic period is estimated from egg diameters to be 22–28 days.

Conus episcopus auctt.

Prior records

Perron and Kohn (1985) noted the egg diameters and estimated the duration of the precompetent planktonic stage of *C. episcopus* from Indonesia and Palau. Further details of these are given under new records below.

New records

1. Pulo Bai, Batu Group, off Sumatra, Indonesia; 26 November 1963. No. 6367. An adult female was collected from sand under a large coral rock in 1 m of water; the egg mass was affixed to the underside of the rock. When examined 6 h after collection, the capsules contained motile trochophore-equivalent embryos. A few uncleaved eggs were present (Fig. 3.7A). They attained early veliger stage by the following day (Fig. 3.7B) but changed little over the next two days (Table 3.1). On the sixth day following collection, veligers were very well developed and very motile within the capsules. They differed from all other known *Conus* veligers in having two narrow, dark purplish-brown pigment bands along the edge of the velar lobes, on either side of the food groove; the latter was also pigmented, but a paler purple. A row of green pigment spots paralleled the bands proximal to them. The foot had brown pigmentation and a very large operculum. Veligers hatched on the tenth day after collection. As the velar lobes expanded, the

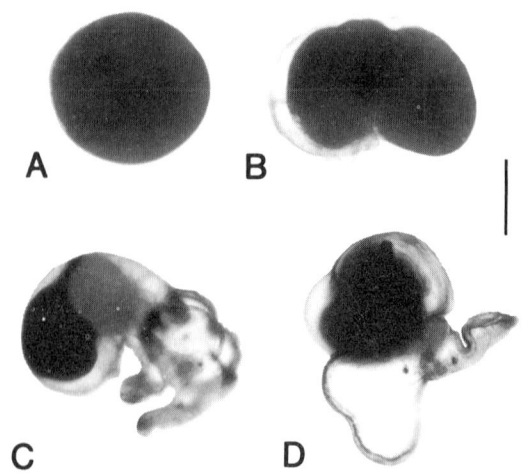

Fig. 3.7 Developmental stages of *Conus episcopus* auctt. No. 6367; Indonesia. (A) Uncleaved egg. (B) Early veliger. (C) Lateral view of hatched veliger. (D) Anterolateral view of hatched veliger, showing expanded, bilobed velar lobe. Scale bar = 0.2 mm.

purplish-brown pigment bands appeared thinner (Fig. 3.7C,D). The eyes were large and otocysts were present. The heart was observed to beat at a rate of 84 min^{-1}. All hatched veligers died within three days with no settlement or metamorphosis.

2. Western barrier reef 0.8 km S of Toagel Mlungui Channel, Palau, Caroline Islands; 9 March 1983. No. FEP-203. An adult female collected crawling among rocks at night later deposited an egg mass in a sea-water tank at MMDC.

3. Ngadarak Reef, Palau, Caroline Islands; 20 June 1983. No. FEP-251. A female was collected with an egg mass under a dead coral slab behind the reef crest.

Both egg masses from Palau contained uncleaved ova which were followed through embryonic and larval development. They hatched as veligers less than 600 μm in shell length after 13 days and, under laboratory conditions, required at least 15 days to attain metamorphic competence. The larvae are planktotrophic.

The Type II egg mass of *C. episcopus* is virtually identical to those of other molluscivorous *Conus* species such as *C. pennaceus* and *C. marmoreus*, with most capsules attached to others rather than to a substratum. Each capsule is about 16–20 × 12–13 mm and contains 700–2300 eggs about 255 μm (Palau) or 400 μm (Indonesia) in diameter. The disparity in egg diameter between the Palau and Indonesia specimens was unexpectedly large; the estimated duration of the precompetent planktonic period of the latter is 6 days.

Conus figulinus Linnaeus

Prior records

Kohn (1960, 1961b) described the Type III egg mass of *C. figulinus*, in which five basal capsules have expanded bases, are devoid of eggs and exit window, and serve to anchor the entire egg mass in the fine sand that is the habitat of this species.

New records

1. 1 km S of Carlatan Harbor, San Fernando, La Union, the Philippines; 25 April 1964. No. JEN-45. An egg mass containing uncleaved ova was collected with a female at a depth of 9 m.

2. N part of Cabugao Bay, 1.5 km NW of Sinait, Ilocos Sur, the Philippines; 28 April 1964. No. JEN-49. An egg mass containing uncleaved ova was collected with a female at a depth of 8 m.

The *C. figulinus* egg mass contains 30–60 capsules 16–22 × 10–13 mm, each with from 3000 to more than 8000 eggs 190–210 μm in diameter.

The duration of the precompetent planktonic period is estimated from egg diameters at 21–24 days.

3. Balgal Beach, N. Queensland, Australia; 12 November 1970. AMS Nos. C.158481, C.158487. Five capsules all attached to 1–3 others from one egg mass, and two capsules from a second egg mass, were collected with four adult females and three males in muddy sand at LWS, by Ian Loch. The embryos were in early cleavages up to the 4-cell stage (255–279 × 279–292 μm).

Conus flavidus Lamarck

Prior record

Perron (1981a,b,c) reared *C. flavidus* from oviposition through hatching, settlement, and metamorphosis in Hawaii. Frank (1969) observed oviposition at Heron Reef, Queensland, Australia, between November and February but gave no further details.

New records

1. Seaward side of Uliga I., Majuro Atoll, Marshall Islands; 9 September 1956. No. 3926. An adult female was collected on limestone reef rock with an algal sand-mat under a large coral rock, with a cluster of 26 egg capsules (Fig. 3.8). Two of these were attached to the substratum; the rest were in rows in a dense cluster on the underside of the rock. The capsules were also attached to each other by

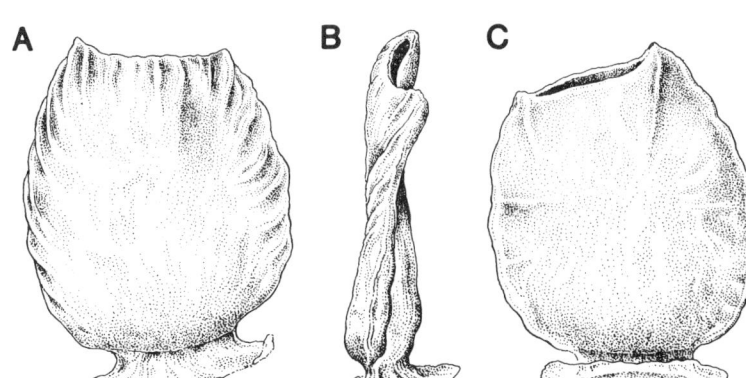

Fig. 3.8 Three views of an egg capsule of *Conus flavidus* Lamarck. No. 3926; Marshall Islands. Scale bar = 5 mm.

confluent, rather large basal plates. One capsule was examined; it contained well-developed veligers about 285 μm in maximum shell dimension, and a few uncleaved eggs 175 μm in diameter.

2. Ngadarak Reef, Palau, Caroline Islands; 10 June 1982. No. FEP-54. A female was observed ovipositing under a dead coral slab on the reef flat. The capsules contained only uncleaved ova.

3. Hio, N of Tuapa, Niue; 7 September 1986. No. BAIU-40. Coll. B. Holthuis. A female with four egg capsules was collected on the underside of a large rock in a tide pool on the reef flat. When fixed, all embryos were at the 4-cell stage, except for one 2-cell stage 273 × 182 μm. Diameter of the uncleaved egg was estimated to be 227 μm.

The egg capsules of *C. flavidus* are 10–12 × 8–10 mm (Fig. 3.8); each contains 2000–2500 eggs 175–230 μm in diameter. They form a Type I egg mass. The duration of the precompetent planktonic period is 23 days (Perron 1981*a*).

Conus frigidus Reeve

Prior record

Kohn (1978) reported briefly on oviposition by *C. frigidus* on the underside of a small coral rock at Pulli Island, Gulf of Manaar, India. The Appendix Table 1 provides additional information on this egg mass.

New records

1. Pulo Stupai, Sanding I., Indonesia; 5 December 1963. No. 6864. A female was observed ovipositing on the underside of a coral rock; three other individuals of *C. frigidus* were under the same rock. When examined 10 h after oviposition, the capsules contained mainly 2- and 4-cell stages. Their development was followed over the next ten days (Table 3.1; Fig. 3.9A–C,F), but the larvae died in the veliger stage without hatching. At 9 days, the rapidly motile veligers had visible otocysts and short tentacles with well-developed eyes. The velar lobes lacked pigment spots.

2. Pulo Stupai, Sanding I., Indonesia; 7 December 1963. Nos. 7141–7147. Five female and two male *C. frigidus* were found under the same coral rock with many egg capsules. These were not collected, but two egg capsules deposited by one of the females after collection contained uncleaved eggs; development was not followed.

3. Pulo Stupai, Sanding I., Indonesia; 7 December 1963. No. 7150. A female was observed ovipositing on the underside of a rock; a second female

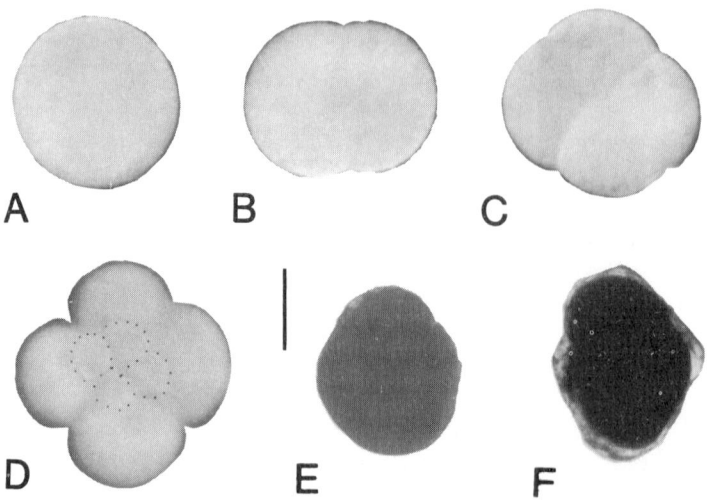

Fig. 3.9 Developmental stages of *Conus frigidus* Reeve. Nos. 6864 (A–C,F) and 7150 (D,E); Indonesia. (A) Uncleaved egg. (B) 2-cell stage. (C) 4-cell stage. (D) 8-cell stage; dots outline first quartet of micromeres. (E) Gastrula. (F) Trochophore-equivalent stage. Scale bar = 100 μm.

and a male *C. frigidus* were under the same rock. Development was followed (Table 3.1; Fig. 3.9D,E) but all embryos died in early stages.

The egg capsules of *C. frigidus* are typically deposited in Type I egg masses on the undersides of coral rocks on reef platforms. Capsules contain 1500–4000 eggs, 190–212 μm in diameter. The duration of the precompetent planktonic period is estimated from these data to be 22–24 days.

Conus furvus Reeve

New records

1. Bokbok I., NE of Culion, Palawan, the Philippines; 18 December 1963. No. JEN-13. A female was collected with an egg mass on a rock in 6 m of water. The capsules contained late veligers averaging 902 μm in maximum shell dimension.

2. Coron Harbor, Palawan, the Philippines; 20 December 1963. No. JEN-59. An adult female and large mass of egg capsules attached to wood and shell fragments were collected in sea grass 2 km W of the pier at a depth of 4 m. The capsules contained only uncleaved ova.

3. 0.5 km NE of Cabilauan I., Coron Harbor, Palawan, the Philippines; 19 December 1963. No. JEN-60. A female with egg mass (Fig. 2.2C) attached to sea grass on top of coral was collected at a depth of 10 m. The egg capsules contained uncleaved ova and early cleavage stages.

The egg masses of *Conus furvus* are of Type II (Fig. 2.2C), as is typical of most molluscivorous species. However, the diet of *C. furvus* is unknown, and its shell colour pattern lacks the 'tented' pigmentation characteristic of most molluscivorous species.

Conus furvus has large eggs (630–700 μm diameter) and the fewest per egg capsule (6–29) of any known Indo-Pacific species of the genus. The largest egg mass (142 capsules) was estimated to contain fewer than 3400 eggs. Another unusual feature is that the eggs and developing larvae are embedded within a dense fibrous material within the capsule. Only preserved material was studied, but the fibrous material appeared to inhibit movement within the capsule at all stages observed. The size of eggs and veligers, which were beginning to develop shell siphons, indicates that *C. furvus* lacks a planktonic larval stage.

Conus geographus Linnaeus

Prior records

Kohn (1961*b*) described and illustrated the egg capsules and larvae of *Conus geographus* from Seychelles. Cruz, Corpuz, and Olivera (1978) described and illustrated mating, spawning, and larval development in *C. geographus* from the Philippines. This species is known to form mating and spawning aggregations of up to 15 individuals, at least in Seychelles (Kohn 1961*b*).

The single egg mass of *C. geographus* studied in detail was of Type I and consisted of 54 egg capsules 26–28 × 18–21 mm, each containing an average of more than 16 000 eggs about 190 μm in diameter. The egg mass thus contained approximately one million eggs. The duration of the precompetent planktonic phase is estimated from egg diameter to be 24 days.

Conus glans Hwass in Bruguière

Prior record

Kohn (1960, 1961*b*) described the egg capsules and larvae of *Conus glans* from Sri Lanka. Duration of the precompetent planktonic stage is estimated from egg diameter to be 3 days, but the prehatching larvae were veliconchas (Kohn 1961*b*: Fig. 10) and a planktonic stage may be lacking.

New record

W barrier reef 1 km S of Toagel Mlungui Channel, Palau, Caroline Islands; 23 April 1963. No. FEP-217. An adult female was found buried deep in dead *Acropora* rubble. It was placed in a sea-water tank at MMDC and was observed ovipositing on 6 May 1983. Hatching began on 20 May 1983. The planktotrophic larvae, 0.7 mm in shell length at hatching, were raised in the laboratory until settlement and crawling behaviour was observed on the ninth day after hatching. All larvae died the following day due to bacterial contamination of the cultures. Metamorphosis was not observed.

Despite the differences in observed egg diameters and likely developmental mode, the egg capsules of both records appeared identical, and we could detect no differences in the shell characteristics of the adults. The egg mass of *C. glans* is of Type I.

Conus imperialis Linnaeus

Prior records

Egg masses and larval development of *Conus imperialis* have been described from Hawaii (Kohn 1961*a*; Perron 1981*a*) and Seychelles (Kohn 1961*b*).

New records

Reef flat behind Ngadarak Reef, Palau, Caroline Islands; 1 July 1982. No. FEP-83. A female and egg mass were collected in 0.5 m of water. It was not possible to follow larval development.

The egg mass of *C. imperialis* is of Type II. The capsules are 13–20 × 9–13 mm; each contains about 1000–4500 eggs, 220–265 μm in diameter. The duration of the precompetent planktonic period is estimated from egg diameters to be 18–22 days.

Conus leopardus (Röding)

Prior records

Egg capsules and larval development of *Conus leopardus* have been described and illustrated from Hawaii and Seychelles (Kohn 1961*a,b*; Perron 1981*a,b*).

New records

1. Lagoon adjacent to Uliga and Jarrej Is., Majuro Atoll, Marshall Islands; 7 September 1956. A cluster of egg capsules was found attached to the underside of a large coral head in an area of sand and rubble at a depth of 1.5 m. They are assigned to *C. leopardus* because of their very large size and close resemblance to egg capsules of this species from the other records listed.

2. Lagoon adjacent to Aomon I., Enewetak Atoll, Marshall Islands; 17 September 1973. An egg mass collected at the site described by Kohn (1981) hatched on 20 September 1973. The foot of the veliger was moderately well developed but appeared to lack an operculum; long stiff cilia were present at the posterior tip of the foot.

3. Malakal Harbor, Palau, Caroline Islands; 25 October 1984. No. FEP-403. A large egg mass was observed under a ledge at the bottom of a channel in the reef at a depth of 3 m. It was possible to collect only a few capsules.

Conus leopardus is the largest Indo-Pacific species in the genus, and its egg capsules are by far the largest of any known species (30–58 × 18–37 mm). Each contains from 3000 to more than 22 000 eggs, fewer than in the smaller capsules of *C. vexillum*. The typical number of capsules in the Type I egg mass is unknown. The largest observed contained 67 (total number of eggs about 750 000) but oviposition was interrupted. Egg capsule size is significantly correlated with adult shell size (Table 3.2). The eggs are moderately small (mean diameter 220 μm), however, and the estimated duration of the precompetent planktonic stage is 21–23 days.

Conus litteratus Linnaeus

New records

1. Pulo Stupai, Sanding I., Indonesia; 7 December 1963. No. 7279. An adult female with egg capsules attached to dead *Acropora* branches was collected; it was not possible to follow larval development.

2. Malakal Harbor, Palau, Caroline Islands; 23 January 1984. No. FEP-313. A female was observed ovipositing on the side of a dead coral boulder on the fringing reef off the MMDC seawall.

3. Cocos Lagoon, Guam, Mariana Islands; 30 July 1957. A female was observed ovipositing on the underside of a coral rock.

The egg mass of *Conus litteratus* is of Type I. The capsules (Fig. 3.10) are quite distinctive and could

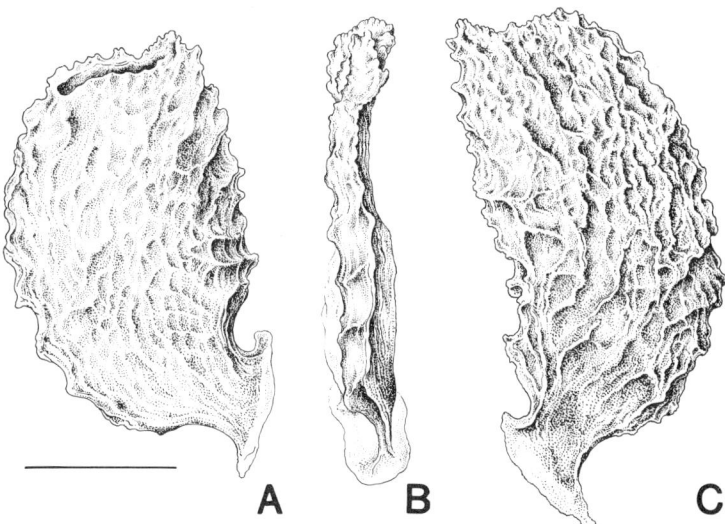

Fig. 3.10 Three views of an egg capsule of *Conus litteratus* Linnaeus. Guam. Scale bar = 1 cm.

hardly be mistaken for those of any other known Indo-Pacific species. The capsule walls bear numerous fine ridges that form a network over the surface and give the capsule an irregular outline.

The capsules are 23–28 × 16–19 mm and contain about 10 000–14 000 eggs 200–225 μm in diameter. The duration of the precompetent planktonic period, estimated from the egg diameter, is 21–23 days.

Conus lividus Hwass in Bruguière

Prior records

Kohn (1961*a,b*) and Perron (1981*a,b,c*) described and illustrated egg capsules and developmental stages of *Conus lividus* from Hawaii, Sri Lanka, Maldives, and Seychelles.

New records

1. Cua Be, Nhatrang, Vietnam; 4 December 1959. Coll. J. Knudsen. An adult female was collected with egg capsules attached to a limestone rock. The capsules contained uncleaved ova.

2. Pulo Penju, NE of Simalur, off Sumatra, Indonesia; 22 November 1963. No. 6151. An adult female was collected on sand under a coral rock to which rows of 9–10 egg capsules were attached. When examined 4.5 h later, the egg capsules contained very motile trochophore-equivalent embryos (Fig. 3.11A). Over the next 3 days these progressed to a well-developed veliger stage (Table 3.1; Fig. 3.11B), but all died prior to hatching.

3. Pulo Melila, Banjak Group, off Sumatra, Indonesia; 23 November 1963. No. 6254. A large female was observed ovipositing on the edge of a sponge and algal mat, in 0.5 m of water. Development was not followed.

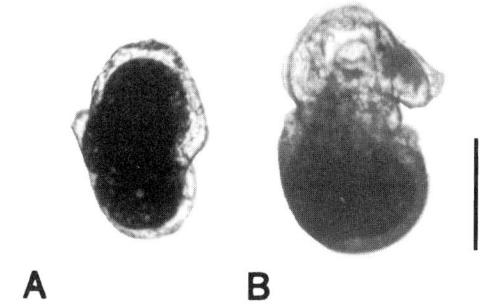

Fig. 3.11 Developmental stages of *Conus lividus* Hwass in Bruguière. No. 6151; Indonesia. (A) Trochophore-equivalent stage. (B) Veliger stage. Scale bar = 100 μm.

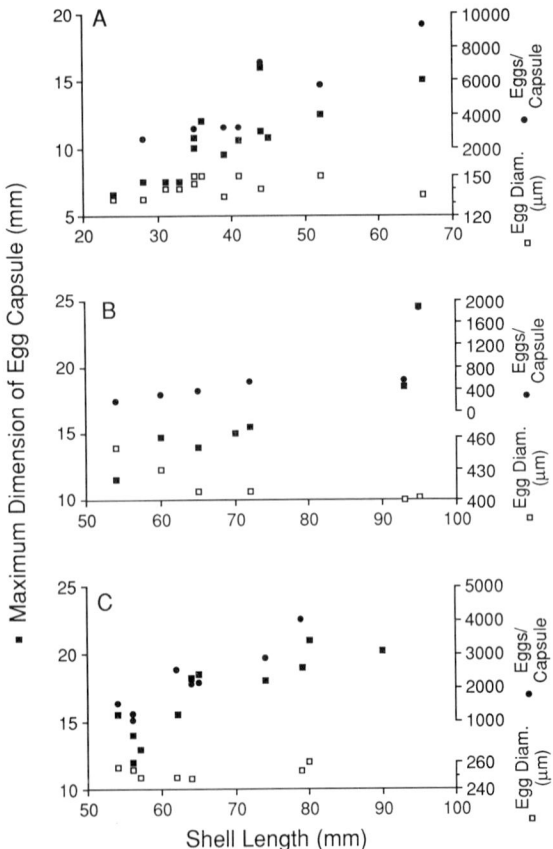

Fig. 3.12 Relationships of egg diameter, egg capsule size, and number of eggs per capsule with body size of mother (as shell length). (A) *Conus lividus* Hwass in Bruguière. (B) *C. marmoreus* Linnaeus. (C) *C. textile* Linnaeus. For correlation coefficients, see Table 4.1.

6. Talin Bay, Batangas, the Philippines; 3 February 1964. No. JEN-35. A single female was collected among coral at a depth of 6 m, with what appeared to be an egg mass deposited by two females. The egg capsules were in two groups that did not overlap in size, 12–14.5 × 11–13 mm (No. JEN-35A) and 10.5–11 × 11 mm (No. JEN-35B). Both contained uncleaved eggs as well as some with the first two cleavage furrows visible. The smaller capsules were assigned to the female present, as they were more appropriate to her body size (Fig. 3.12A).

7. Nikao, Rarotonga, Cook Is.; 10 October 1984. No. BRAR-69. Coll. G. Paulay. A female was observed ovipositing under a rock on the fringing reef. When fixed, most embryos were in the 4-cell stage.

The records of *Conus lividus* egg masses are sufficiently numerous to examine the relationship between adult size, egg capsule, and egg size (Fig. 3.12A; Table 3.2). Larger females lay larger egg capsules but do not produce larger eggs. The average number of eggs per capsule is also significantly correlated with body size. The *C. lividus* egg mass is of Type I.

The duration of the precompetent planktonic phase of *C. lividus* is estimated from egg diameters to be 28–29 days, although larvae cultured by Perron (1981*a*) in Hawaii did not settle until 50 days after hatching.

Conus magus Linnaeus

New records

4. Pulo Stupai, Sanding I., off Sumatra, Indonesia; 7 December 1963. No. 7258. An adult female was observed ovipositing on the underside of a coral rock in 0.8 m of water. Embryos examined 10 h later were in the 4-cell stage. They failed to develop normally, and all died within 4 days.

5. Salong, Calapan, Mindoro, the Philippines; 14 April 1964. No. JEN-12. A small cluster of recently deposited egg capsules containing uncleaved ova was found with an adult female at a depth of 8 m.

1. Pulo Melila, Banjak Islands, Indonesia; 23 November 1963. No. 6238. A female was collected with a cluster of 80 freshly deposited egg capsules. They were retained in sea-water for 18 days, but most embryos were dead in the veliger stage on the 15th day and all had died by the 18th day after oviposition (Table 3.1). Capsules fixed 13 h after collection contained uncleaved eggs. By 2½ days they had reached gastrula stage. Slowly motile early veligers, with eyes and small velar lobes bearing a few green pigment spots were

SPECIES ACCOUNTS

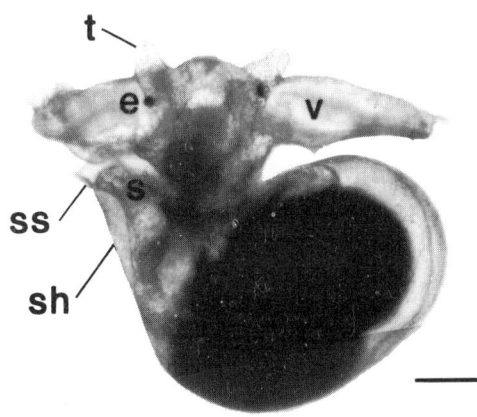

Fig. 3.13 Prehatching veliger of *Conus magus* Linnaeus. No. 6238; Indonesia. e, Eye; s, siphon; ss, shell siphon; sh, shell; t, tentacle; v, velar lobe. Scale bar = 100 μm.

present at 6½ days. At 8 days, otocysts were visible and the foot was large and flat with a small operculum. At 12 days, the veligers were well developed, with a clear metachronal beat of the velar cilia, and with a shell siphon. The large heart beat at a rate of 120 min^{-1}. Cilia on the sole of the foot were observed to reject particles. At 15 days (Fig. 3.13), dorsal pigmentation on the well-developed foot was observed, but there was no operculum. Tentacles were also well developed. When next examined 3 days later, all veligers were dead, and none had hatched.

2. Pulo Melila, Banjak Islands, Indonesia; 23 November 1963. No. 6240. A female was observed ovipositing in a clump of *Halimeda opuntia*. Capsules fixed 13 h after collection contained uncleaved eggs. By 2½ days they were only at third cleavage. All embryos died without developing past gastrula stage (Table 3.1).

3. Lagoon, Ponape, Caroline Islands; 24 August 1956. No. 3284. An egg mass of 30 mostly empty capsules was found attached to the underside of a rock; a female *C. magus* was partly buried in sand under the same rock. The intact capsules contained embryos whose development appeared to have been arrested in the early veliger stage; they measured about 500 μm in maximum dimension. This egg mass is assigned only tentatively to the female associated with it and is excluded from the statistical analyses reported below.

4. Calatagan, Batangas, the Philippines; 9 February 1964. No. JEN-8. An egg mass containing late veliger larvae was collected with an adult female in 4 m of water on a rock and coral bottom.

5. Mouth of Coloconto Bay, Batangas, the Philippines; 20 March 1964. No. JEN-10. An egg mass containing mainly late veliger to veliconcha stage larvae was collected in 5 m of water on a coral head on a sand bottom. The veliconcha shells were 1.16 mm long, and their diameter, probably equivalent to Protoconch I, was 0.5 mm.

6. 2 km E of Barrio Castanas, Quezon, the Philippines; 15 March 1964. No. JEN-21. Two females were found together, one with a mass of egg capsules on the shell, at a depth of 9 m on grey sand. Capsules containing uncleaved eggs and advanced veliconchas were present.

7. 1 km S of San Diego Pt., Talin Bay, Batangas, the Philippines; 3 February 1964. No. JEN-34. An egg mass containing mainly early to late veligers was collected in 10 m on a sand and coral bottom.

8. Pulo, Santa Cruz Is., Marinduque, the Philippines; 6 April 1964. No. JEN-46. An egg mass collected with an adult female in 4 m of water on a sand bottom contained late trochophores and early veligers.

9. Villa Carmen, Batangas, the Philippines; 1 March 1964. No. JEN-53. An egg mass collected with an adult female was attached to *Acropora* rubble under a rock in 6 m of water. Most of the 44 capsules were attached directly to the substratum with confluent bases, but six were attached only to other capsules. They contained only uncleaved ova.

10. 3 km S of San Diego, Talin Bay, Batangas, the Philippines; 3 February 1964. No. JEN-54. An egg mass collected with an adult female in 10 m of water on a coral and sand bottom contained trochophores and early veligers.

11. 2 km S of Talin Pt., Talin Bay, Batangas, the Philippines; 5 February 1964. No. JEN-55. An egg mass collected with an adult female in 10 m of water on a coral and sand bottom contained uncleaved eggs and 2-cell stages.

12. Grande Is., W of Capulaan Bay, Quezon, the Philippines; 22 March 1964. No. JEN-58. An egg mass collected with an adult female in 6 m of water on a coral and sand bottom contained uncleaved eggs and embryos in which the first two cleavages had occurred.

The *Conus magus* egg mass is of Type I but a small proportion of capsules are sometimes attached to other capsules. Average egg diameters in the 11 samples ranged from 499 to 580 μm. This size range, and the presence of advanced veliconchas in some of the older egg masses indicate that this species lacks a planktonic larval stage.

Larger females had larger egg capsules and more eggs per capsule, but the correlations were not statistically significant (Table 3.2).

Conus marmoreus Linnaeus

New records

1. Lagoon, Ponape, Caroline Islands; 26 August 1956. No. 3347. An egg mass attached to the underside of coral rock was collected; the female was partly buried in sand adjacent to the rock. The egg capsules contained mainly veliger larvae.

2,3. Uliga I, Majuro Atoll, Marshall Islands; 4 September 1956. Nos. 3600, 3610. Two females were observed ovipositing on the undersides of dead coral rocks in 1 m of water in the lagoon. No. 3600 had deposited only two capsules before oviposition was interrupted. Capsules from both females contained uncleaved ova when examined.

4. Reef flat behind Ngadarak Reef, Malakal Harbor, Palau, Caroline Islands; 20 May 1982. No. FEP-25. The female deposited an egg mass in the laboratory on 27 May 1982.

5. Western barrier reef 3 km N of Aulong Channel, Palau, Caroline Islands; 5 August 1982. No. FEP-101. The female deposited an egg mass in the laboratory on 22 August 1982. Swimming veligers hatched 16 days after oviposition.

6. Maniuayan Is., Marinduque, the Philippines; 8 April 1964. No. JEN-16. A female with an egg mass was collected at a depth of 6 m.

7. Dauis, Bohol, the Philippines; 4 May 1964. No. JEN-19. A female with egg mass was collected on a rocky and sandy bottom at a depth of 4 m. The capsules contained mainly 4-cell stages.

The egg mass of *C. marmoreus* is a robust Type II structure with many interconnected capsules that measure 11–19 × 7–14 mm (Figs. 3.14, 3.15). Each contains about 200–600 eggs 400–450 μm in diameter.

Larval development in *C. marmoreus* is rather similar to that of *C. consors*. The very large hatchlings (0.84 mm) from Palau had to swim and feed for at least 8 days before attaining metamorphic competence. The fully developed larvae (1.4 mm in shell length) metamorphosed either

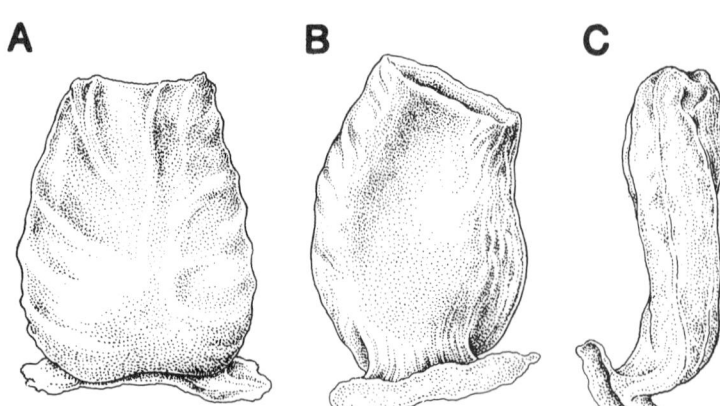

Fig. 3.14 Three views of an egg capsule of *Conus marmoreus* Linnaeus. No. 3610; Marshall Is. Scale bar = 5 mm.

SPECIES ACCOUNTS

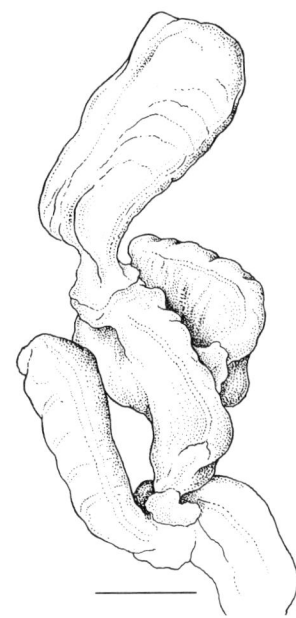

Fig. 3.15 Part of an egg mass of *Conus marmoreus* Linnaeus. No. 3610; Marshall Is., showing detail of attachment of egg capsules to others in a Type II egg mass. Scale bar = 5 mm.

spontaneously in the fibreglass rearing tanks or when placed on small stones or coral chips. Rates of successful metamorphosis were 36–72 per cent. Egg diameters of *C. marmoreus* from several geographic regions indicate a precompetent planktonic period of 2–6 days. Egg capsule size and number of eggs per capsule were significantly correlated with adult size, and egg diameter was inversely correlated with adult size (Fig. 3.12B, Table 3.2).

Conus miles Linnaeus

New record

Barrier reef at Oa, Ponape, Caroline Islands; 27 August 1956. No. 3444. A female was collected with a cluster of egg capsules in a few cm of water on rough coral substratum on the reef crest. Oviposition was not observed, and the egg capsules were extremely large relative to the size of the adult, but they contained uncleaved ova. They are tentatively assigned to *C. miles*. The larger *C. distans* occurs in the same habitat, but its eggs are much smaller (148 vs. 228 μm) than those in the present egg mass.

Conus miliaris Hwass in Bruguière

Prior records

Kohn (1961b) described the eggs and capsules of *C. miliaris* from the Indian Ocean. Embryonic development was not followed. Frank (1969) observed oviposition at Heron Reef, Queensland, Australia, in February but gave no further details.

New record

Goh Sindarar Nua, Thailand; 6 November 1963. No. 6031. A female was completely buried in sand under a small dead coral rock to which the egg capsules were attached in rows of up to eight, in 0.3 m of water. A capsule fixed on the day of collection contained uncleaved eggs. One day later, the pear-shaped, ciliated embryos were probably in gastrula stage, as were those fixed the following day. Trochophore-equivalent embryos were present 3.5 days after collection; capsules fixed 5.7 and 6.7 days after collection contained early veligers with eyes, very small velar lobes, a well-formed shell. When released from the capsule, they were able to locomote along the bottom of the dish with the velar cilia but did not swim. The only pigment present was a brown strip ventrally between the velar lobes. Eight days after collection, the velar lobes were still rather small, but 6–8 widely separated green pigment spots were now present on each lobe. Otocysts and eyes were present, but tentacles had not yet developed. The small foot bore a large operculum. One day later, veligers were well developed, able to swim well when released, and with brown pigment along the growing edge of the shell. Hatching occurred on the following day. The strongly swimming larvae had a black stripe along the edge of the velum in addition to well-developed green pigment spots. Most veligers died after swimming for about a day, and by the third day after hatching all had died. Table 3.1 summarizes the course of development.

The egg mass of *C. miliaris* is of Type I. The capsules are only 5–11 × 6–9 mm; each contains about 1000 eggs 150–200 μm in diameter. These hatch as very small (0.30 mm) veliger larvae, comparable in size to those of the closely related species *C. abbreviatus* and *C. coronatus*. The duration of the precompetent planktonic stage is estimated from egg diameters at 23–27 days.

Conus moreleti Crosse

Prior record

Kohn (1961*b*) described and illustrated the egg capsules and embryonic development of *C. moreleti* from Seychelles. The egg mass is of Type I. Egg capsules 10–11 × 8 mm contain about 2000 eggs 150 μm in diameter, indicating a precompetent planktonic period of 28 days.

Conus obscurus Sowerby

Prior record

Perron (1981*a*) briefly noted egg size (147 μm), prehatching period (12–13 days), hatching size (0.26 mm), and size at settling (2.10 mm) of *C. obscurus* in Hawaii. Duration of the precompetent planktonic period is estimated from egg diameter to be 28 days.

Conus omaria Hwass in Bruguière

New records

1,2. Wading I., W of Viti Levu, Fiji; May 1962. No. 10247. A. Jennings collected two females with egg masses.

3. Kayangel Atoll, Palau, Caroline Islands; 12 June 1982. No. FEP-61. A female was collected in sand under a rock and returned to the laboratory, where an egg mass was deposited 24 June 1982. The egg capsules contained uncleaved eggs when first observed; they developed to hatching veliger larvae in 14 days.

4. Igedelukes Reef, Palau, Caroline Islands; 29 March 1983. No. FEP-210. A female was observed ovipositing under a coral rock in 0.5 m of water. Development to hatching veligers occurred over the next 14 days.

As in other molluscivorous *Conus* species, the egg mass of *C. omaria* is of Type II. The capsules are 12–16 × 8–15 mm and contain about 500–1000 eggs 330–340 μm in diameter.

Larvae of both egg masses from Palau grew and metamorphosed 12 days after hatching. The duration of the precompetent planktonic period in the Fiji sample is estimated from egg diameter to be 11 days.

Conus pennaceus Born

Prior records

Ostergaard (1950) first reported on the egg capsules and development of *C. pennaceus* (referred to by him as *C. omaria*). He noted the small number of very large eggs and the absence of a planktonic veliger larval stage. Kohn (1961*a*) reported on ten additional egg masses from Hawaii, confirming and supplementing Ostergaard's results. Kohn (1961*b*) described an egg mass from the Maldive Islands. Perron (1981*a*) also reared *C. pennaceus* through metamorphosis in Hawaii, and Perron (1981*b*, 1982) and Perron and Corpuz (1982) studied several aspects of reproductive energetics of this species.

New records

1. Hare Island, off Tuticorin, India; 18 February 1968. No. 7351. Kohn (1978) gave a brief account and photograph of the egg mass of a female on the underside of a large rock at a depth of 1 m. The capsules contained mainly well-developed veligers and a few uncleaved eggs about 630 μm in diameter. One day after collection, eyes, tentacles, and statocysts were visible. The velum was bi-lobed; the anterior lobes were larger. The velum bore a single row of small green pigment spots and rather short cilia. The large foot also had short cilia. At 3½ days after collection, veligers remained similar in appearance but were more active. When released from the capsule, they glided along the bottom of the dish, apparently propelled only by the cilia of the velar lobes. They appeared unable to swim up off the bottom. None was seen to crawl on the foot, which appeared no larger than it was 2 days earlier. It was necessary to terminate the observations 5½ days after collection, at which time the veligers were generally similar in appearance and activity as at 3½ days. The foot appeared to have enlarged somewhat, but all movement remained due to velar ciliary beating.

2. Hare Island, of Tuticorin, India; 18 February 1968. No. 7352. A female was in the process of affixing her sixth egg capsule to the underside of a rock in 1 m of water. A capsule examined the following day contained mainly 4-cell-stage embryos. A few uncleaved eggs about 650 μm in diameter were also present. Development did not progress beyond this stage, and the capsules were fixed 4 days after collection.

3. Pulo Bai, Batu Group, off Sumatra, Indonesia; 29 September 1963. No. 6318. A female with egg mass was collected. The capsules contained well-developed veligers with dense green pigment spots on the velum, prominent eyes, and foot with operculum (Fig. 3.16). A few uncleaved eggs 390 μm in diameter were also present.

The striking differences in egg diameter and apparent developmental mode between populations of *C. pennaceus* from different geographic regions cannot be adequately explained at present. Mean egg diameters in the samples examined are 390 μm (Maldives), 407 μm (Indonesia), 490 μm (Hawaii), and 641 μm (India). Larvae from the first two areas are predicted to have precompetent planktonic periods of 7 and 5 days, respectively; those from the latter areas are expected to have a very brief (Hawaii) or no (India) planktonic stage. All egg masses were of Type II, but egg capsule form differed. Those from Indonesia conform with prior descriptions of *C. pennaceus* egg capsules from Hawaii and Maldives (Ostergaard 1950; Kohn 1961*a,b*). The lateral edges are convex and slightly scalloped and the surface bears indistinct ridges. The surface of egg capsules from India (No. 7351) is unusually smooth (Kohn 1978: Fig. 5). Two low ridges extend basally from the corners of the exit window, but there is no other surface ornamentation. The capsules are also more quadrangular, with only slightly convex edges. These differences may indicate within-species variation, or *C. pennaceus* as presently understood may actually be a complex of closely related species. Among the correlations with adult body size examined over all the samples included as *C. pennaceus*, only egg capsule size was significant (Table 3.2).

Conus pertusus Hwass in Bruguière

Prior record

Perron (1981*a*) briefly noted egg size (132 μm), prehatching period (11 days), hatching size (0.28 mm) and the unusually large size at settling (2.15 mm) of *C. pertusus* in Hawaii. Duration of the precompetent planktonic period is estimated from egg diameter to be 29 days.

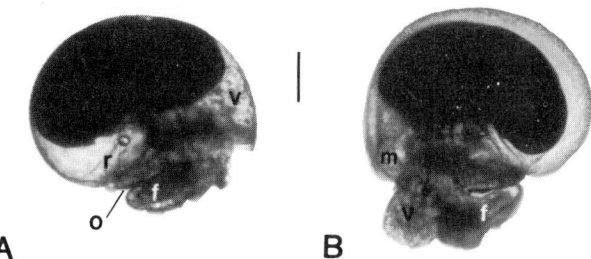

Fig. 3.16 Prehatching veligers of *Conus pennaceus* Born. No. 6318; Indonesia. (A) Right lateral view. (B) Left anterolateral view. f, Foot; m, mantle, with opening to mantle cavity; o, operculum; r, retractor muscle; v, velar lobe, retracted into mantle cavity in A, extended in B; y, yolk. Scale bar = 100 μm.

Conus planorbis Born

New record

Bauan, Batangas, the Philippines; 3 April 1964. No. JEN-22. A female and Type I egg mass attached to sunken wood on a sand bottom were collected at a depth of 4 m. The capsules contained uncleaved ova and dissociated early cleavage stages.

The duration of the precompetent planktonic period is estimated from egg diameter to be 22 days.

Conus pulicarius Hwass in Bruguière

Prior records

Perron (1981a: Table 1) reported briefly on the development of *C. pulicarius* in Hawaii. Huish (1978) briefly noted egg size and number of eggs per capsule in a *C. pulicarius* egg mass from Lizard Island, Queensland, Australia.

New record

Lagoon side of Japtan I., Enewetak Atoll, Marshall Islands; 1 September 1956. No. 3489. The Type I egg mass consisted of 21 capsules attached to the underside of a coral rock, under which the adult female was completely buried in coarse sand at a depth of 1 m. In contrast to those of most *Conus* species, the egg capsules were broader than high (8–9 × 10–11 mm) (Fig. 3.17). Many additional egg masses were observed on the undersides of nearby rocks, but no other adults were observed.

C. pulicarius is estimated to have a precompetent planktonic period of 26–28 days, based on egg diameter.

Conus quercinus [Lightfoot]

Prior records

Kohn (1961a) described and illustrated the reproductive habits, egg masses, and development of *C. quercinus* from oviposition to hatching as veligers 15–16 days later. Perron (1981a,b,c) cultured larvae from oviposition to hatching and through settlement and metamorphosis 30 days later.

New records

1. Salong, Calapan, Mindoro, the Philippines; 14 April 1964. No. JEN-6. Two females were collected with one egg mass at a depth of 6 m. Most eggs were uncleaved or had one or two visible cleavage furrows.

2. Little Balateros, Pto. Galera, Mindoro, the Philippines; 16 April 1964. No. JEN-9. Two females with a very large egg mass of about 250 capsules were found under a rock in 6 m of water. The egg mass is possibly from both females. It contained uncleaved eggs and early cleavage stages.

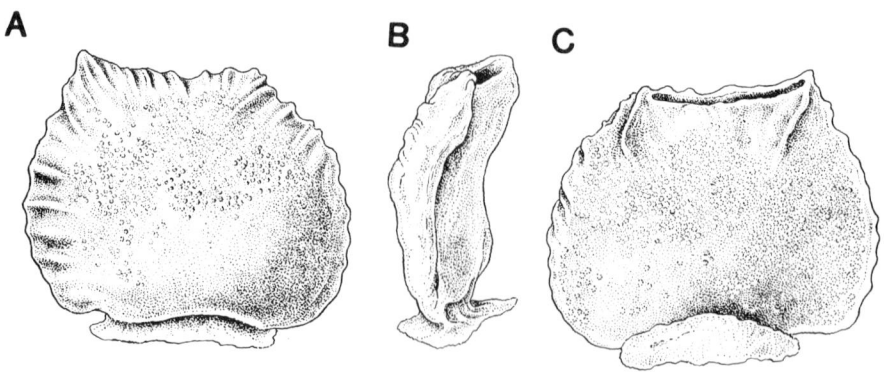

Fig. 3.17 Three views of an egg capsule of *Conus pulicarius* Hwass in Bruguière. No. 3489; Marshall Is. Scale bar = 5 mm.

3. 1 km N of Taytayen Pt., Batangas, the Philippines; 20 January 1964. No. JEN-17. A male and female, both bearing egg capsules attached to their shells, were collected on sand with an egg mass attached to the test of a clypeasteroid echinoid. The egg capsules comprising the egg mass occurred in three distinct groups. They varied more in size than is usual for the egg capsules of a single female (Appendix), so it is uncertain whether they comprise one or more egg masses. All, however, ranged from uncleaved eggs to 4-cell embryos.

4. Bauan, Batangas, the Philippines; 3 April 1964. No. JEN-38. A female and egg mass were collected at a depth of 6 m on sand. The egg capsules contained mainly uncleaved eggs, with a few 2-cell stages.

Conus quercinus is one of a few species in the genus known to form mating and spawning aggregations. These have long been known to occur, typically in the month of February, at Ahuolaka (Sand) Island, Kaneohe Bay, Oahu, in 1–2 m (A. H. Banner, pers. com. to AJK, 1954), but few quantitative data exist. We sampled this site at approximately monthly intervals in 1956, with the following results:

1 January: 4 males.
9 February: 4 males, 5 females, 3 sex not determined. Three heterosexual pairs (animals within 15 cm), one with an egg mass. Three additional egg masses on sponge and *Acanthophora*.
6 March: 5 males, 1 female, 1 sex not determined. No egg capsules.
21 March; 27 April; 12 June: No adults.

However, in 1954, E.C. Jones (pers. com.) collected 10 adults (6 males, 4 females) at the same site on 5 May. He did not note the presence of any egg capsules.

The egg mass of *C. quercinus* is of Type I but with confluent basal plates. Presumably this feature maintains the integrity of the egg mass when it is affixed to a narrow substratum such as a sponge or the alga *Acanthophora*. The capsules range in height from less than 1 cm to more than 2 cm; they are often affixed to plants on a sand bottom. Most contain 1500–3000 eggs, and an egg mass may approach and perhaps exceed a half million eggs.

The precompetent planktonic period of *C. quercinus* in Hawaii was observed to be 30 days (Perron 1981a). The specimens from the Philippines have somewhat smaller eggs (mean diameter 169 μm vs. 190 μm in the Hawaiian sample reared through metamorphosis). The estimated duration of the precompetent planktonic period based on the Philippines samples is 25–26 days.

Conus rattus Hwass in Bruguière

Prior records

Ostergaard (1950) described and illustrated egg capsules and early embryonic development of *C. rattus* (as *C. tahitensis rattus* Hwass) in Hawaii. Kohn (1961a) described the veliger larva, and Taylor (1975) recorded settlement and metamorphosis, also in Hawaii. Egg capsules of *C. rattus* have also been described from Sri Lanka and Seychelles (Kohn 1961b).

New record

Reef flat behind Ngatarak Reef, Palau, Caroline Islands; 4 May 1984. No. FEP-365. A female was observed ovipositing under a dead coral slab. It was not possible to follow development.

The egg capsules of *C. rattus* are 8–15 × 6–14 mm; they are deposited in Type I clusters on the undersides of coral rocks on reef platforms. Each capsule contains about 2000–6000 eggs as small as 125 μm in diameter, the smallest eggs known in the genus. Correlations of spawn characteristics with adult size were not significant (Table 3.2). Duration of the precompetent planktonic period is estimated from egg diameter to be 25–30 days.

Conus retifer Menke

New record

Pulo Penju, off Sumatra, Indonesia; 22 November 1963. No. 6157. A female and Type II egg mass of

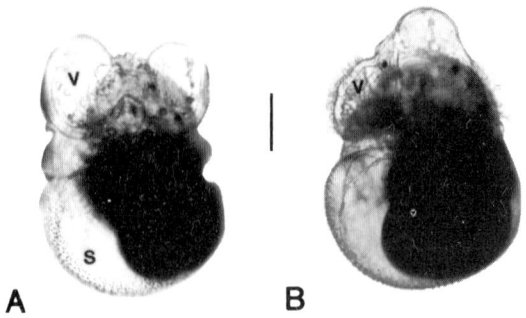

Fig. 3.18 Prehatching veligers of *Conus retifer* Menke. No. 6157; Indonesia. (A) Ventral view. (B) Ventrolateral view. s, Shell, with brown granules on surface; v, velar lobe. Scale bar = 100 μm.

91 capsules 9 × 6–7 mm were in a crevice on the underside of a coral rock in 0.8 m of water. Observation of the egg capsules 4 h after collection revealed actively swimming veligers with well-developed foot and operculum, pale green pigment spots on the velar lobes, and eyes, retractor muscles, and statocysts visible (Fig. 3.18). It was not possible to retain the egg mass alive until hatching. The duration of the precompetent planktonic period, estimated from diameters of the few undeveloped eggs present (253 μm), is 19 days.

Conus sponsalis nanus Sowerby

Prior record

Perron (1981*a*) briefly noted egg size (135 μm), prehatching period (11 days), hatching size (0.22 mm) and size at settling (1.45 mm) of *C. sponsalis nanus* in Hawaii. Duration of the precompetent planktonic period is estimated from egg diameter to be 29 days.

Conus stercusmuscarum Linnaeus

New records

1. W of Lacay Lacay Pt., Cagayan, the Philippines; 4 May 1964. No. JEN-1. A female with egg mass was collected from sand at a depth of 6 m. The capsules contained uncleaved eggs and a few early cleavage stages.

2. Muelle Pt., Galera, Mindoro, the Philippines; 17 April 1964. No. JEN-41. A female with egg mass was collected at a depth of 6 m. The egg capsules contained mainly early veligers.

3. Mompog I., Marinduque, the Philippines; 8 April 1964. No. JEN-47. A female with egg mass was collected from a rock at a depth of 6 m. The egg capsules contained uncleaved ova.

4. W barrier reef, Palau, Caroline Islands; 15 October 1982. No. FEP-115. A female was observed ovipositing under dead coral on the shallow reef. It was not possible to follow development.

Conus stercusmuscarum egg masses are of Type I and consist of about 60 egg capsules 15–20 × 12–18 mm. Each contains about 1000–2700 eggs 235–290 μm in diameter. This gives an estimate of the duration of the precompetent planktonic stage of 15–20 days.

Conus stramineus Lamarck

New record

0.1 km S of Laiya, Sigayan Bay, Batangas, the Philippines; 13 March 1964. No. JEN-11. Two adults were collected with a Type I egg mass on a sand bottom at a depth of 7 m. The egg capsules (9–11 × 6.5–81.5 mm) were deposited in rows on a solitary tunicate, the tunic of which incorporated a tightly adhering layer of sand grains. Each capsule contained only 31–56 veligers about 635 μm in maximum dimension.

Based on the 527 μm diameter of the few uncleaved eggs present, *C. stramineus* is predicted to lack a planktonic larval stage.

Conus striatellus Link

New record

Reef flat behind Ngadarak Reef, Palau, Caroline Islands; 14 April 1984. No. FEP-347. A female was observed ovipositing under a rock. When examined in the laboratory, the freshly laid capsules (14–15 × 10 mm) contained about 1800 uncleaved ova

193 μm in diameter. They were retained in a flowing sea-water system until planktotrophic veligers hatched 11 days later. Settlement and metamorphosis were not observed. The duration of the precompetent planktonic stage is estimated from egg diameter to be 24 days.

Conus striatus Linnaeus

Prior record

Perron (1981*a,c*) reared *C. striatus* from oviposition to hatching 16 days later as planktotrophic veliger larvae. These settled and metamorphosed 20 days after hatching.

New records

1. Heron Reef, Great Barrier Reefs, Australia; 24 December 1962. No. 10324. Dr Isobel Bennett collected a female with egg mass on the reef platform.

2. Heron Reef, Great Barrier Reefs, Australia; 9 February 1975. A female was observed ovipositing on the underside of a large flat coral rock in 0.3 m of water on the reef platform. Five of the 29 egg capsules were removed and returned to the laboratory. When examined 4 h later, two contained uncleaved ova, and one each contained eggs in first cleavage division, 2-cell stage, and 4-cell stage. On the third day after oviposition, the capsules contained slowly motile, rotating embryos that were probably in blastula stage. On the sixth day, the embryos ranged from early to well-developed trochophore-equivalent stage and were more rapidly motile. They attained well-developed veliger stage by the ninth day. Pale green pigment spots were observed on the velar lobes on the eleventh day. On the following day they were more numerous, and the very large velar cilia were easily visible. Hatching occurred on the 14th day after oviposition, but all hatched veliger larvae died 4 days later. The course of development is summarized in Table 3.1.

3. Agana Bay, Guam, Mariana Islands; 29 July 1957. A cluster of 10 large, hatched and empty egg capsules was found attached to the underside of a rock. Four adult *C. striatus*, an unusually common species at this location, were on the sand under the rock.

4. Punta Nasugbu, Batangas, the Philippines; 12 April 1964. No. JEN-29. A female with egg mass was collected at a depth of 6 m. The capsules contained mostly early veliger stages.

5. Salong, Calapan, Mindoro, the Philippines; 14 April 1964. No. JEN-40. A female and egg capsules were collected at a depth of 8 m under a rock. The capsule contents were too poorly preserved to analyse.

6. Sand flat behind Ngadarak Reef, Palau, Caroline Is.; 14 April 1984. No. FEP-348. A female and egg mass were collected under a rock. The capsules contained mostly veliger stages.

The egg capsules of *C. striatus* are 16−30 × 14−25 mm (Fig. 3.19) and contain about 1700−6500 eggs 235−267 μm in diameter. They are deposited in Type I clusters on the undersides of rocks. The duration of the precompetent planktonic phase in *C. striatus* is estimated from egg diameters of samples from Australia, the Philippines and Palau to be 18−20 days, consistent with the observed duration in Hawaii.

Conus striolatus Kiener

New record

Augulpelu Reef, Palau, Caroline Islands; 21 October 1982. No. FEP-107. A female was observed ovipositing on the underside of a dead coral head. Some of the egg capsules were retained in a flowing sea-water system until hatching was observed 13 days later. The duration of the precompetent planktonic period is estimated from egg diameter to be 21 days.

Conus suratensis Hwass in Bruguière

New record

One km SW of San Fernando Beacon Pt., San Fernando Bay, La Union, the Philippines; 23 April 1964. No. JEN-52. Two adult females and a male were collected with a Type I mass of 93 capsules

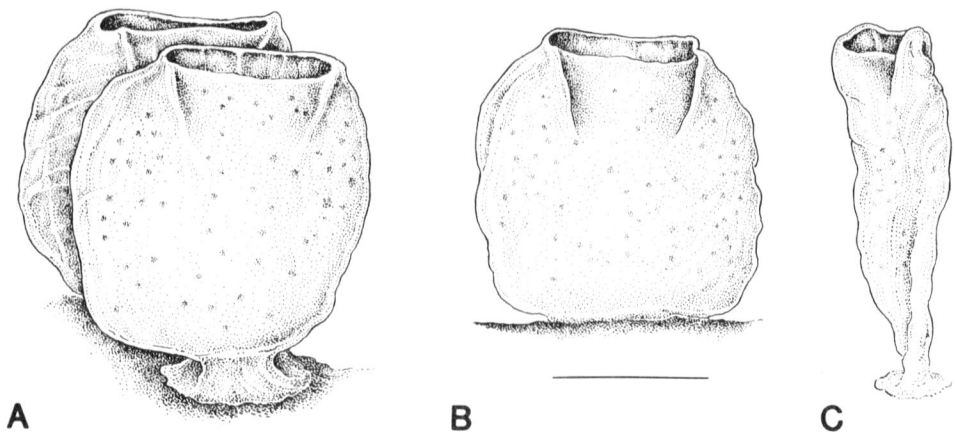

Fig. 3.19 Three views of an egg capsule of *Conus striatus* Linnaeus. Guam. Scale bar = 1 cm.

(16–17 × 14–15 mm) attached to filamentous algae on a sand bottom at a depth of 8 m. Each capsule contained about 4400 eggs 207 μm in diameter.

Based on the uncleaved eggs contained in these capsules, the duration of the precompetent planktonic phase in *C. suratensis* is estimated to be 23 days.

Conus tessulatus Born

Prior record

Thorson (1940a) described and illustrated egg capsules and developmental stages of this species from the Persian Gulf. From his reported egg diameter of 250 μm, we estimate the duration of the precompetent planktonic stage to be 19 days. The identification of these capsules with *C. tessulatus* must be considered tentative, because of the large size of the capsules relative to the size of most *C. tessulatus* adults.

Conus textile Linnaeus

Prior records

Huish (1978) briefly noted egg diameter and number of eggs per capsule in an egg mass collected in 18 m 'under fallen plate coral, between bommies', at Wheeler Reef, Queensland, Australia. Perron (1980, 1981a,b) cultured *Conus textile* through hatching, settlement and metamorphosis in the laboratory in Hawaii and provided details on larval feeding and growth.

New records

1. Pu Renget Basar Reef, Singapore; 21 October 1963. No. 5925. An egg mass was collected on the underside of a *Porites lutea* head. The capsules contained veligers with shells about 365 μm in diameter and 415 μm long. These hatched from the capsules 4 days after collection. At hatching, the velar lobes were very large and mobile, with a broad row of pale opaque yellow pigment spots around the edge and scattered spots over the surface of the lobe, as illustrated by Perron (1980: Fig. 1A) for this species in Hawaii. The veligers swam for about three days, after which they all died.

2. Airport Reef, Goh Phuket, Thailand; 17 September 1963. No. 6126. An adult female had oviposited five capsules on the underside of a coral rock at a depth of 2 m. Another female and two males were also present under the same rock. Most of the eggs were uncleaved 6.5 h after oviposition. Three days later, embryos were in gastrula stage

(245–270 × 205–220 μm), but it was not possible to follow further development.

3. Airport Reef, Goh Phuket, Thailand; 17 September 1963. No. 6129. A female with egg capsules affixed to the underside of a coral rock was collected at a depth of 2 m. Embryos were in blastula to gastrula stages on the day of collection. Gastrulas were about 268 × 232 μm. Development was followed until hatching 10 days later (Table 3.1), but the free-swimming veligers survived for only 2 days.

4. Pulo Bai, Batu Group, off Sumatra, Indonesia; 27 November 1963. No. 6413. A female with egg mass was found under a large coral head in 0.8 m of water. On the day of collection, the capsules contained well-developed veligers with prominent green pigment spots on the velar lobes, eyes, and foot with operculum. Hatching began the following day and continued for at least one day. Hatched veligers survived for only 2 days.

5. Pulo, Santa Cruz Is., Marinduque, the Philippines; 6 April 1964. No. JEN-20. Two females with an egg mass were collected on a rock at a depth of 1.5 m. The capsules contained well-developed veligers.

6. Dauis, Bohol, the Philippines; 4 May 1964. No. JEN-27. A female and egg mass were collected from rock and sand at a depth of 5 m. The capsules contained well-developed veligers.

7. 1 km NW of Rabon, Lingayen Gulf, Pangasinan, the Philippines; 16 April 1964. No. JEN-28. A female and egg mass were collected under a rock at a depth of 1.5 m. The capsules contained uncleaved eggs and some with the first two cleavage furrows.

8,9. Salong, Calapan, Mindoro, the Philippines; 14 April 1964. Nos. JEN-30 and JEN-39. Two females with egg masses were collected under rocks at a depth of 6 m. The egg capsules of No. JEN-30 contained uncleaved eggs; those of No. JEN-39 contained early veliger stages.

10. W of Damortis, Lingayen Gulf, Pangasinan, the Philippines; 17 April 1964. A female and egg mass were collected from rocks among sea grass at a depth of 6 m. The egg capsules contained uncleaved ova.

11. Man Friday Resort, near Korolevu, Viti Levu, Fiji; 12 November 1988. No. BFIJ 326. Coll. B. V. Holthuis. A female with egg capsules affixed to the underside of a large rock resting on sand and rubble at a depth of less than 1 m. The egg capsules contained embryos in gastrula stage and were fixed on the day of collection.

C. textile spawns compact Type II masses of up to 62 egg capsules 12–21 × 8–19 mm, many of which are affixed to earlier laid capsules. The egg mass is usually deposited on the underside of a coral rock. Each capsule contains about 1000–4000 eggs 250–280 μm in diameter. These hatch as veligers 11–17 days after oviposition. Egg capsule size, number of eggs per capsule, and number of eggs per egg mass are directly related to parent body size, but egg diameter is independent of body size of the mother (Fig. 3.12C; Table 3.2). In Hawaii, *C. textile* has a minimum planktonic period in the laboratory of 16 days (Perron 1980). Estimates based on egg diameters from other geographic regions are 16–21 days.

Conus thalassiarchus Sowerby

New records

1,2. El Nido Bay, Bacuit, Palawan, the Philippines; 8 May 1962. No. JEN-24; JEN-no number. Two egg masses and adults were collected on a sand bottom in 4–6 m. The egg masses were attached to clumps of different species of *Halimeda*. Egg capsules in one contained mainly uncleaved eggs with some 2- and 4-cell stages; egg capsules in the other contained mainly 4-cell stages.

3. Coron Harbor, Calamian Group, Palawan, the Philippines; 20 December 1963. No. JEN-25. A female and cluster of mainly empty egg capsules were collected on the side of a rock on a sand and coral bottom at a depth of 6 m. A few capsules contained well-developed veligers about 0.81 mm in maximum dimension.

The egg mass of *C. thalassiarchus* is of Type II. The egg capsules are small (7.5–9.5 × 5–6.5 mm) relative to parent body size. They contain few (20–50) very large (600 μm) eggs, and the largest number of eggs estimated in an entire egg mass was

4100. The very large egg diameters indicate that *C. thalassiarchus* lacks a planktonic larval phase.

Conus varius Linnaeus

Prior records

Kohn (1961b) described and illustrated egg capsules and developmental stages from three egg masses of *C. varius* from Seychelles.

New records

1. Salong, Calapan, Mindoro, the Philippines; 14 April 1964. No. JEN-31. A female with egg capsules was collected under a rock at a depth of 8 m. Most capsules contained well-developed veligers about 265 μm in maximum dimension.

2. Reef flat behind Ngadarak Reef, Palau, Caroline Is.; 15 June 1983. No. FEP-242. A female and egg mass were collected under a rock in 0.5 m of water. Most embryos were near hatching stage, but several uncleaved ova were present.

The egg capsules of *C. varius* are 8–13 × 6–10 mm. They are deposited in a Type I cluster with confluent bases and contain about 2000 eggs 160–195 μm in diameter. The duration of the precompetent planktonic period is estimated from egg diameters at 24–27 days.

Conus vexillum Gmelin

Prior records

Ostergaard (1950) and Kohn (1961b) reported on and illustrated egg capsules and some aspects of embryonic development in *C. vexillum* from Hawaii (reported as *C. sumatrensis* by Ostergaard 1950) and Seychelles, respectively. Perron (1981a,b) provided additional information on development and size at settlement.

Conus vexillum produces a Type I egg mass of large capsules (20–31 × 13–22 μm). The number of eggs per capsule may exceed 50 000 and the number per egg mass may exceed 1.5 million, more than in any other known species of *Conus*. Hatching veligers measure only 0.25 mm and grow to the unusually large size of 2.0 mm prior to settlement, during a minimum planktonic period estimated from egg diameter to be 28 days.

Conus victoriae Reeve

New record

1. Port Warrender, Admiralty Gulf, Western Australia; 31 July 1976. Western Australian Museum No. 288–91. Coll. Fred E. Wells. A Type II egg mass of 132 capsules was collected among intertidal rocks with six adults of undetermined sex. The capsules were 11–14 × 7–9 mm and each contained 52–75 uncleaved eggs 631 μm in diameter. The large egg size indicates that *C. victoriae* has nonplanktonic development.

Conus virgo Linnaeus

Prior records

Kohn (1961b) reported on the egg capsules and embryos of *C. virgo* from Sri Lanka and Seychelles.

New records

1. Nagiba (E side of Balayan Bay), Batangas, the Philippines; 22 February 1964. No. JEN-36. An adult with egg mass was collected at a depth of 8 m on a coral bottom.

2. Lagoon, Jarrej I., Majuro Atoll, Marshall Islands; 10 September 1956. No. 3944. A female was observed ovipositing on the underside of a coral rock in 1 m of water.

The egg capsules of *C. virgo* are 12–32 × 12–22 mm. They are deposited in a Type I cluster with confluent bases and contain 8000–18 000 eggs 170–212 μm in diameter. An egg mass may contain more than 700 000 eggs. The duration of the precompetent planktonic phase in *C. virgo* is estimated from egg diameters at 21–26 days.

Conus vitulinus Hwass in Bruguière

Prior record

Kohn (1961a) described and illustrated egg capsules and developmental stages of *C. vitulinus* from Hawaii. Capsules in a Type I egg mass were 23 × 16–17 mm and eggs 225–250 μm in diameter. Veligers, 400 μm in maximum dimension, hatched 14–16 days after oviposition. Duration of the precompetent planktonic stage is estimated from egg diameters to be 19–21 days.

4 RELATIONSHIPS AMONG ASPECTS OF REPRODUCTION AND LIFE HISTORY

This chapter addresses the questions of how the important parameters of reproduction and early life history are interrelated, both within and between species of *Conus*. They include adult body size, egg size, egg capsule size and number of eggs per capsule, fecundity, developmental mode, and dispersal ability. Here we synthesize and interpret the basic data presented in Chapter 3 and summarized in tabular form in the Appendix.

METHODS

For statistical analyses of the data documenting these relationships, we have generally followed Sokal and Rohlf (1981). We estimated the relationships among the various life history characteristics and between these and properties of adults by regression analysis. For these bivariate analyses of continuous variables, Model II regression is more appropriate than the more familiar Model I or ordinary least squares (OLS) regression, because in our data sets, the magnitudes of the independent variables used (e.g. egg size, egg capsule size, adult body size) are not controlled by the investigators. Thus both dependent and independent variables are subject to error because of inherent variability, as well as measurement error.

In addition, Model II regression is more appropriate when the primary goal is explanation, that is to determine the functional relationship, not merely to predict values of Y from X, and this applies to most of our analyses. As is typical of levels of biological organization at and above the organism (e.g. Kohn 1989), the processes studied here are sufficiently complex that the distinction between explanation and prediction is important (Levins 1970; Wimsatt 1982). The Model II regression used here is the reduced major axis (RMA) regression, for the reasons given by Sokal and Rohlf (1981) and McArdle (1988); see also LaBarbera (1989).

We generally present the results of both RMA and Model I (OLS) regressions, for three reasons. OLS is more familiar and more widely understood; it is appropriate in cases where prediction of values of dependent variables is desired, and it permits a test of the significance of the observed association, that is of the difference between the observed regression slope and a slope of zero. Such a test is inappropriate in RMA regression. The slope of the RMA regression is always higher than that of the corresponding OLS regression; the difference is proportional to the correlation coefficient. For both models, we indicate the slopes as well as the 95 per cent confidence limits, with the RMA confidence limits calculated by the method of Jolicoeur and Mosimann (McArdle 1988).

Although generally a secondary goal in this study, prediction is particularly relevant in two important cases: the regression of minimum planktonic or precompent larval period on egg diameter, and the potential use of the regression of egg diameter on larval shell size to predict the latter from the protoconchs of adult shells.

EGG SIZE

We treat egg size first, because as noted below it appears to be the most important controlling or operational variable on other aspects of reproduction and development. Egg sizes of 61 Indo-Pacific species of *Conus* are now known (Appendix Table 1). Their average diameters range from 125 to 825 μm, with a median of 225 μm. Plotted as a frequency distribution, the mean within-species values have a unimodal pattern with positive skew (Fig. 4.1).

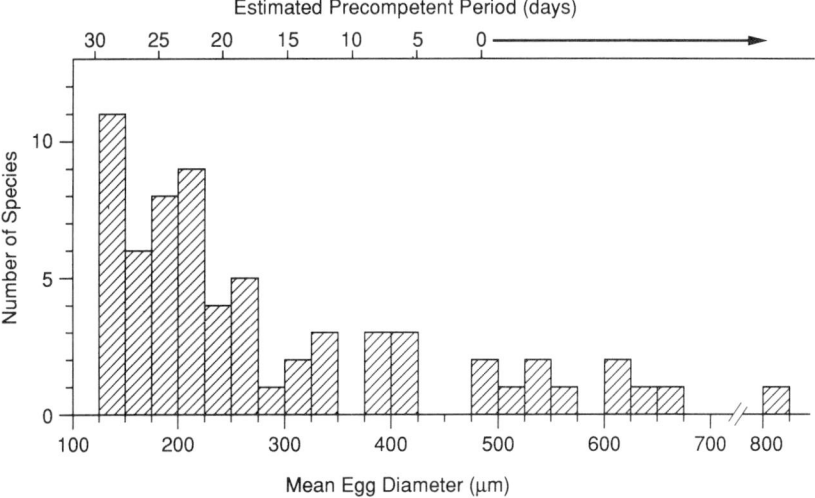

Fig. 4.1 Egg and frequency distributions of 61 Indo-Pacific *Conus* species. Because of large intraspecific differences between geographically distant populations of some species, the following are entered separately: *C. eburneus*: Australia, Enewetak; *C. episcopus*: Indonesia, Palau; *C. pennaceus*: Hawaii, India, Indonesia, Maldives. $N = 66$. Median diameter = 224 μm. Complete data are given in Appendix Table 1.

On the basis of data then available, Kohn (1961*b*) concluded that the distribution of egg size in Indo-Pacific *Conus* species had a bimodal pattern, with most species 100–250 μm in diameter and a few, 350–500 μm. We now know that the absence of species with egg diameters between 250 and 350 μm was an artifact of small sample size. In the more complete sample reported here, the largest gaps are between 665 and 825 μm, 420 and 490 μm, 345 and 390 μm, and 275 and 310 μm (Fig. 4.1).

The eggs of a female *Conus* are quite uniform in size. The frequency distributions of coefficients of variation (CV) of egg diameters within egg masses (Appendix Fig. 1) indicate that the largest value was 10, 90 per cent of the values were less than 6, and the median value was less than 3. The egg sizes of different females of the same species are somewhat more variable. For species represented by more than one egg mass, Appendix Fig. 1 shows the frequency distribution of coefficients of variation calculated from mean egg diameters of each brood, i.e. within-species, among-individual values. These are generally somewhat higher (median CV = 5.5, maximum CV = 20) than the within-brood coefficients. The three values of CV > 16 represent the species with large interpopulational differences in egg size that are represented by more than one entry in Figs. 4.1 and 4.2.

EGG NUMBER

The number of eggs in a single capsule ranged over nearly four orders of magnitude, from 6–12 (mean = 9.5) in *Conus furvus* to a maximum of more than 50 000 in *C. vexillum*. *C. leopardus*, the largest species in the genus, has more than 20 000 eggs per capsule. Even within a single egg mass, there is a considerable range in number of eggs per capsule. Because eggs were counted in only a few capsules from each egg mass, some of the large apparent differences between individuals of the same species may be attributable to within-individual variation. The frequency distribution of the mean number of eggs per capsule by species (Fig. 4.2) is unimodal, negatively skewed, and leptokurtic, rather than lognormal.

The total number of eggs per egg mass is necessarily less accurately estimated. In addition to the high variability in egg number among capsules of a single egg mass noted above, interruption of oviposition by the investigator results in underestimation of the number of capsules per mass. Our data for 44 species (Fig. 4.3) include six cases in which oviposition was probably interrupted, although a large number of capsules had been deposited. As in the case of number of eggs per capsule, the frequency distribution of number of eggs per egg mass is unimodal, negatively skewed, and leptokurtic (Fig. 4.3).

Egg diameter and number of eggs are strongly and inversely related, but because egg number is also a function of egg capsule size and capsule size is in turn a function of adult body size, we discuss these relationships in the next section.

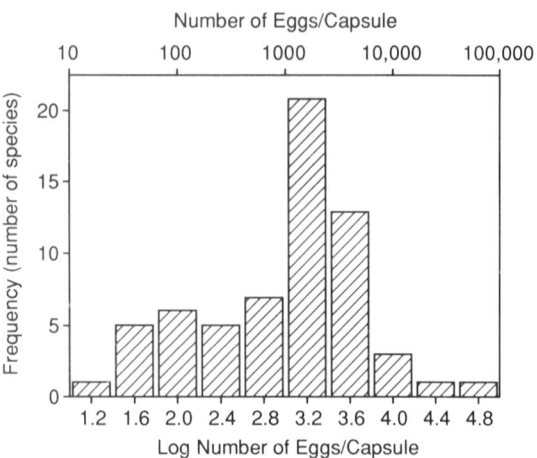

Fig. 4.2 Frequency distribution of mean number of eggs per capsule among 58 Indo-Pacific *Conus* species. Data for three species are entered separately by geographic region as in Fig. 4.1. $N = 63$. Mean = 2768 (log mean = 3.4). Median = 1260 (log median = 3.1). Complete data are given in Appendix Table 1.

ADULT SIZE AND REPRODUCTIVE CHARACTERISTICS

We did not detect any pattern of relationship between egg size and size of mother in *Conus*, either intra- or interspecifically. For the 11 species with observations on at least five egg masses from different mothers, Table 3.2 indicates the generally very low correlations of egg size and adult size (see also Figs. 3.2, 3.12). Females often mature sexually at roughly half their adult size, resulting in a very large size range of reproductive females in several species (24–66 mm in *C. lividus*, 54–95 mm in *C. marmoreus* (Fig. 3.12), and 80–140 mm in *C. leopardus*). Despite this, their egg diameters vary little. We may conclude from this that egg size is a rather constant species characteristic.

Among species, a plot of egg diameter against female size indicates considerable scatter and no clear trend in the data ($r = 0.06$) (Fig. 4.4A). Only the upper right portion of the scatterplot is devoid of points. This means that no species with shells 60 mm or longer had eggs larger than 420 μm in diameter. Figure 4.4A plots the actual size of mothers as the independent variable. The result is essentially similar if the maximum adult shell length for each species is used ($r < 0.03$) (Fig. 4.4B).

Other reproductive attributes do scale with adult size within species, however. Egg capsule size was positively and significantly related to adult body size in nine species; the correlation coefficients were positive in the other four but were not significant at the 0.05 level due to the small sample sizes. Similarly, the number of eggs per capsule increased with body size of the mother, significantly so in five species. In five others the correlation coefficients are strongly positive but not significant at the 0.05 level due to the small sample sizes (Table 3.2).

Interspecific comparisons also indicated a highly significant, positive, linear relationship between adult size and egg capsule size (Fig. 4.5). As noted above, however, egg diameter is not related to body size; thus egg size and capsule size are not significantly correlated ($r = 0.19$).

In interspecific comparisons, species with larger egg capsules generally deposited more eggs in them (Fig. 4.6A). Although the regression accounted for

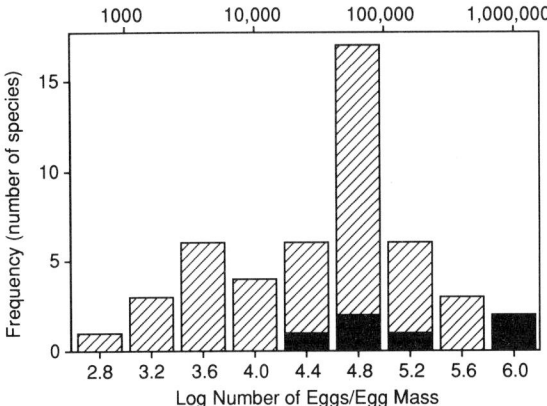

Fig. 4.3 Frequency distribution of mean total number of eggs per egg mass among 44 Indo-Pacific *Conus* species. Absolute numbers of eggs per egg mass are indicated on the scale at top of graph. Solid histograms indicate underestimates due to interruption of female during oviposition. Because of large intraspecific differences between geographically distant populations of some species, the following are entered separately: *C. eburneus*: Australia, Enewetak; *C. pennaceus*: Hawaii, India, Indonesia, Maldives. $N = 48$. Mean = 124 000 (log mean = 5.1). Median = 51 000 (log median = 4.7). Complete data are given in Appendix Table 1.

only about one fifth of the variance, the slope is significant at the 0.001 level. The exponent of the regression is 1.95, indicating that egg number scales with the square of the linear dimension, or with the area of egg capsule. Because the capsules are shaped like flattened flasks, this result supports the hypothesis that the amount of respiratory surface area that the capsule walls provide for the embryos limits the number of eggs per capsule.

The number of eggs per capsule is inversely related to egg diameter (Fig. 4.6B). A negative power function of the form

egg number = a(egg diameter)^b

fits the data much more closely ($r = 0.85$) than the linear function

egg number = $a + b$(egg diameter)
($r = 0.34$; $P < 0.01$).

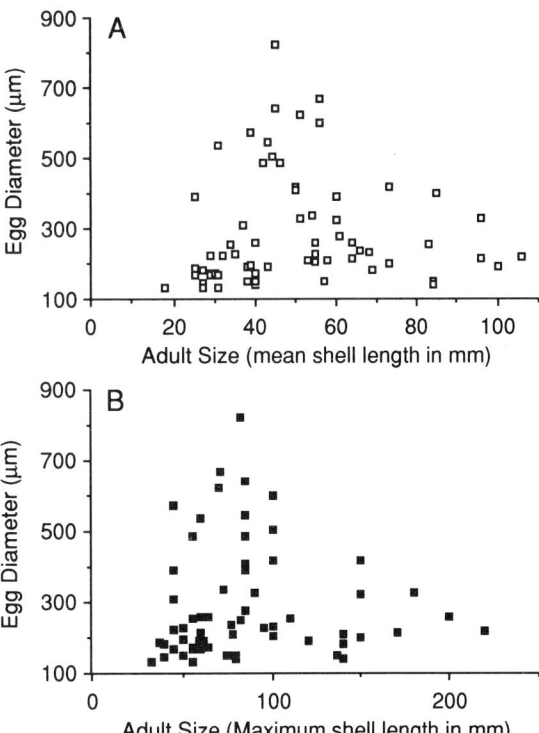

Fig. 4.4 (A) Scattergram of egg diameter vs. shell length of mother in 61 Indo-Pacific *Conus* species. Data for three species are entered separately as in Fig. 4.1. (B) Scattergram of egg diameter vs. maximum shell length of the same species as in (A). Data on maximum shell lengths from Röckel *et al.* (1993).

Because the value of r is so high, the OLS and RMA regressions are very similar. Fecundity, estimated as total number of eggs per egg mass, is also inversely related to egg diameter (Fig. 4.6C). This relation is also better described by the power function ($r = 0.79$) than by the linear function ($r = 0.35$; $P < 0.05$). In both cases the exponent of egg diameter is close to -3, indicating that egg number scales linearly with egg volume or mass rather than with egg diameter. The somewhat lower slopes in Fig. 4.6B than in Fig. 4.6C may indicate that the number of eggs per capsule increases more rapidly with decreasing egg size than does the number of eggs per egg mass. In this way species allocating

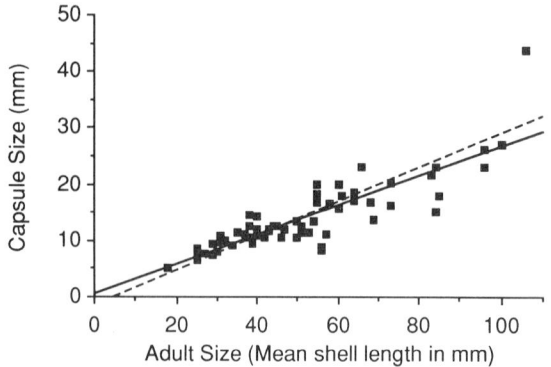

Fig. 4.5 Interspecific relationship of mean egg capsule size (maximum linear dimension) and mean body size of mother (shell length). RMA (dashed line): $Y = 0.31X - 1.50$ (95% confidence limit, CL = 0.27, 0.35). OLS (solid line): $Y = 0.26X + 0.48$ (95% CL = 0.22, 0.30; $r^2 = 0.73$; $P < 0.001$). ($N = 64$).

relatively less energy to capsule material may also produce disproportionately more eggs. Their eggs are smaller as well as more numerous, and more are packaged in each capsule. As will be shown below, embryos of these species also spend less time developing in the egg capsule prior to hatching as veliger larvae (Fig. 4.7).

ESTIMATING EGG DIAMETER FROM ADULT SHELLS

Data on egg diameter are relatively difficult and time-consuming to obtain. Our data set, collected over more than 30 years, provides this information for perhaps 20 per cent of the *Conus* species in the Indo-Pacific region. For this reason, we have examined the relationship between egg diameter and size of the protoconch or larval shell. The protoconch is carried for life at the apex of the adult shell, although it is typically abraded in the majority of shallow water, reef-dwelling *Conus* species that live in rather turbulent environments. We were able to determine both egg diameter and maximum diameter of the first whorl of the shell of 16 species. The OLS and RMA regressions are nearly identical (Fig. 4.8), and egg size may be reliably estimated

for any *Conus* species for which adult shells with well-preserved protoconchs are available. This technique refines and complements earlier methods that related gross larval shell morphology to developmental mode in gastropods generally (Thorson 1950; Shuto 1974; Robertson 1976; Jablonski and Lutz 1983).

EGG SIZE, PREHATCHING PERIOD, AND PRECOMPETENT PERIOD

Perron (1981*b*: Fig. 3) reported a very close relationship between egg diameter and prehatching development period for ten Hawaiian species of *Conus*. The regression $Y = 0.03X + 8.34$, where Y = time in days to hatching and X = egg diameter in μm, explained 97 per cent of the variance. Data are now available for 34 broods, representing 25 species, of Indo-Pacific species that were followed from oviposition to hatching (Fig. 4.7). Although the scatter increased, the regression remained very similar (OLS: $Y = 0.02X + 8.90$; $r^2 = 0.46$; $P = 0.001$; RMA: $Y = 0.03X + 6.93$; for 95 per cent confidence limits, see Fig. 4.7).

We have previously reported the very tight regression of length of the precompetent planktonic period (T_P) on egg diameter (OLS: $Y = 40.10 - 0.08X$; $r^2 = 0.92$; $P < 0.001$), based on 11 species reared through metamorphosis in the laboratory under uniform culture conditions (Perron and Kohn 1985). We have no further data to add to that account, but make a slight correction to the regression (Fig. 4.7). Figure 4.7 indicates the frequency distribution of estimated T_P as well as of egg size, based on the above regression. It shows the results for all species for which T_P was determined in the laboratory, and estimated T_P for those species whose prehatching durations were determined. About two-thirds (39/61) of all species are estimated to have a T_P of 3–4 weeks, while 10 species (about 15 per cent) are considered to lack a planktonic phase. Because larger eggs have longer intracapsular development times, length of the precompetent planktonic period is inversely related to prehatching duration ($r = -0.53$; $P < 0.002$),

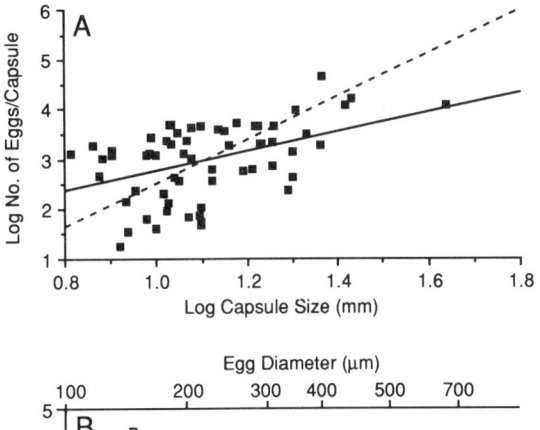
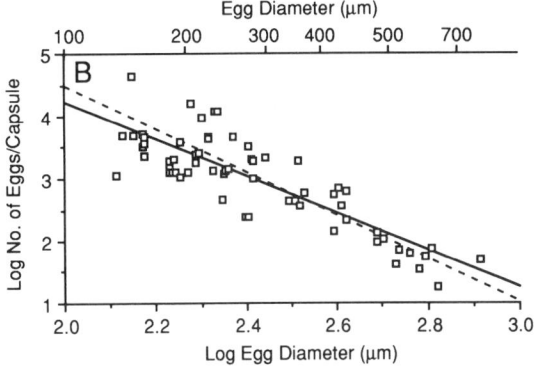
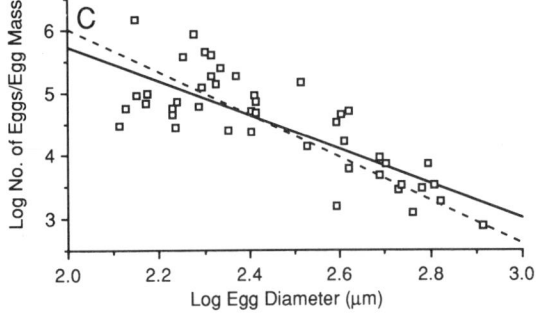

Fig. 4.6 Interspecific comparisons of reproductive attributes in Indo-Pacific *Conus*. (A) Number of eggs per capsule as a function of capsule size (maximum dimension). The RMA power function is $Y = 63.1X^{4.4}$ (log Y = 4.4logX − 1.88; 95% CL = 3.5, 5.5). The OLS power function is $Y = 6.8X^{1.9}$ (log Y = 1.9logX + 0.8; 95% CL = 0.9, 3.0; $r^2 = 0.21$. $P < 0.001$). ($N = 63$). (B) Number of eggs per capsule as a function of egg diameter. The OLS linear regression $Y = 7.2 \times 10^3 - 14.6X$ has $r^2 = 0.11$. The OLS power function $Y = 1.6 \times 10^{10}X^{-3.0}$ (log Y = −3.0logX + 10.2; solid line) fits the data much more closely (95% CL = −2.5, −3.5; $r^2 = 0.72$; $P = 0.0001$). The RMA power function (dashed line) is $Y = 3.1 \times 10^{11}X^{-3.5}$ (log Y = −3.5logX + 11.5; 95% CL = −3.1, −4.0). ($N = 62$). (C) Fecundity or number of eggs per egg mass as a function of egg diameter. The OLS linear regression $Y = 3.1 \times 10^5 - 590X$ has $r^2 = 0.12$. The OLS power function $Y = 1.1 \times 10^{11}X^{-2.7}$ (log Y = −2.7logX + 11.1; solid line) fits the data much more closely ($r^2 = 0.63$; $P = 0.0001$; 95% CL = −2.1, −3.3). The RMA power function (dashed line) is $Y = 6.0 \times 10^{12}X^{-3.4}$ (log Y = −3.4logX + 12.8; 95% CL = −3.0, −4.2). ($N = 47$).

and the two lines in Fig. 4.7 converge as T_P approaches zero with increasing egg size. That is, total development time from oviposition to metamorphic competence, indicated by the total height of the histograms in Fig. 4.7, decreases significantly with increasing egg diameter, up to about 470 μm (ANCOVA with duration as covariate: F = 185, $P < 0.001$). Development time likely increases again for species with larger eggs, but present data are not adequate to demonstrate this.

For all species, the mean prehatching time is 14.6 days (SD = 4.1 days; CV = 28), and mean precompetent planktonic period is 17.4 days (SD = 10.1 days; CV = 58). Excluding those species without a planktonic larval phase, mean T_P is 20.9 days (SD = 7.0 days; CV = 33; $N = 55$). Mean total development time to metamorphic competence in species for which both prehatching and precompetent durations could be determined is 32.1 days (SD = 10.9 days; CV = 34; $N = 29$).

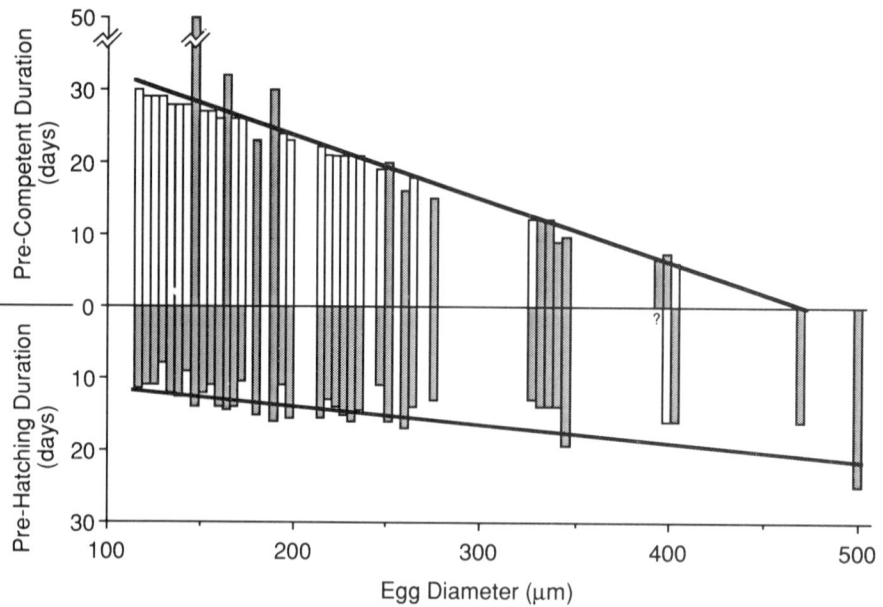

Fig. 4.7 Interspecific relationships between durations of prehatching (intracapsular) development and precompetent planktonic period in Indo-Pacific *Conus*. Each histogram represents minimum total development time from oviposition to metamorphic competence in a single egg mass, partitioned into prehatching (below zero line) and precompetent (above zero line), plotted against egg diameter. At the scale of the graph the RMA and OLS regression lines are indistinguishable for both the prehatching and precompetent periods vs. egg diameter. For prehatching period, RMA: $Y = 0.03X + 6.93$ (95% CL = 0.026, 0.042). OLS: $Y = 0.02X + 8.90$ (95% CL = 0.013, 0.029); $r^2 = 0.46$; $P < 0.001$). ($N = 37$; 21 species are represented once and 8 are represented twice). The regression line shown for precompetent period vs. egg diameter is the experimentally determined relationship for 11 species reared through metamorphosis in Hawaii and Palau (Fig. 5.3): RMA: $Y = -0.09X + 42.18$ (95% CL = $-0.073, -0.113$). OLS: $Y = -0.09X + 40.78$ (95% CL = $-0.07, 0.11$; $r^2 = 0.92$; $P = 0.001$). This regression is used to estimate the precompetent periods of the other species indicated. Divergence of the regression lines to the left indicates that total development time to competence is longer in species with smaller eggs. Stippled histograms, observed values; open histograms, estimated values.

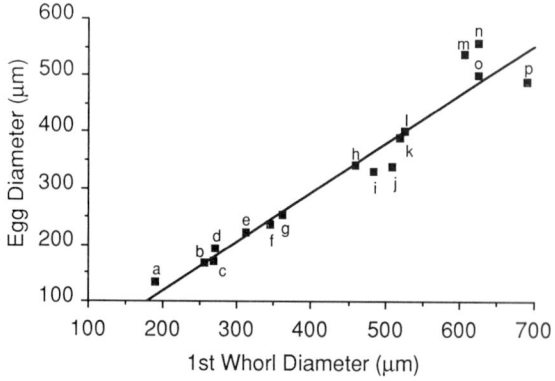

Fig. 4.8 Regression of egg diameter on first protoconch whorl diameter in Indo-Pacific *Conus* species. The protoconch measurement used is maximum diameter of the first whorl (Hansen, 1980). (a) *C. rattus*; (b) *C. varius*; (c) *C. abbreviatus*; (d) *C. striatellus*; (e) *C. coffeae*; (f) *C. striatus*; (g) *C. textile*; (h) *C. glans*; (i) *C. omaria*; (j) *C. ammiralis*; (k) *C. consors*; (l) *C. marmoreus*; (m) *C. stramineus*; (n) *C. magus*; (o) *C. pennaceus*; (p) *C. cinereus*. At the scale of the graph the RMA and OLS regression lines are indistinguishable. RMA: $Y = 0.89X - 65.21$ (95% CL = 0.79, 1.04). OLS: $Y = 0.87X - 53.7$ (95% CL = 0.74, 0.99; $r^2 = 0.94$; $P = 0.001$).

SIZE AT HATCHING AND AT METAMORPHOSIS

Perron (1981a) showed that size at hatching in *Conus* (Y in μm) is reliably predicted by egg diameter (X in μm) based on 19 species from Hawaii, according to the OLS regression $Y = 2.67X - 167$ ($r^2 = 0.96$). Incorporating data now available from 23 additional species from other parts of the IP region alters the regression somewhat ($Y = 2.42X - 144$; $r^2 = 0.93$) (Fig. 4.9).

In marked contrast to size at hatching, size at settling was not related to egg size in the sample of Hawaiian species Perron (1981a) analysed. Rather, the seven species with planktonic larvae that were reared through metamorphosis in the laboratory all settled at a shell length of 1.12–1.51 mm, although their size at hatching varied by a factor of 3 (0.25–0.76 mm). The non-planktonic species *C. pennaceus* hatches and settles with a shell length of 1.1–1.3 ($\bar{x} = 1.25$) mm. The present study adds data for five additional species (*C. consors*, *C. episcopus*, *C. glans*, *C. marmoreus*, and *C. omaria*). All have planktonic larvae and shell lengths at settlement between 1.32 and 1.47 mm. Veliger larvae captured from the plankton in Hawaii and reared individually in test-tubes through metamorphosis (Taylor 1975) have larger shell lengths at settlement (1.25–2.15 mm; $\bar{x} = 1.71$; $N = 7$) (Perron 1981a). Analysis of all 20 species with known size at settlement confirms the result from the smaller Hawaiian sample; length of the veliconcha shell at settlement is not significantly correlated with egg diameter (Fig. 4.10).

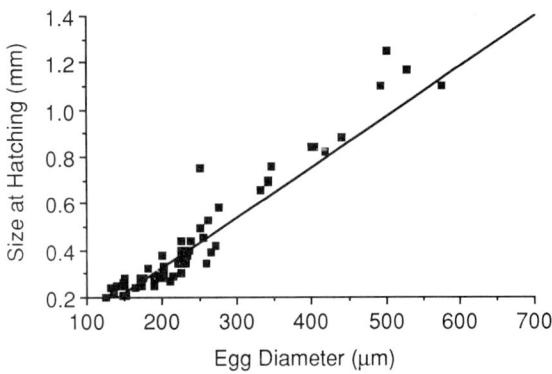

Fig. 4.9 Relationship between size at hatching and egg diameter in 40 Indo-Pacific *Conus* species. Eight species are represented twice and three are represented thrice, giving $N = 54$. At the scale of the graph the RMA and OLS regression lines are indistinguishable. RMA: $Y = 2.49X - 163$ (95% CL = 2.31, 2.69). OLS: $Y = 2.42X - 144$ (95% CL = 2.23, 2.60; $r^2 = 0.93$; $P = 0.0001$).

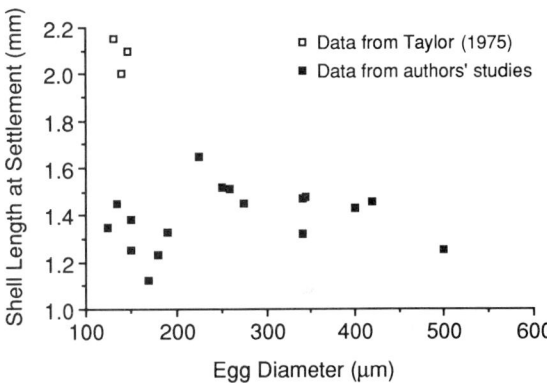

Fig. 4.10 Length of the veliconcha shell of *Conus* species at settlement plotted against egg diameter. ($N = 20$). The correlation coefficient ($r = -0.30$) is not significant at the 0.05 level. The data from Taylor (1975) are based on veliger larvae collected from the plankton.

PROTOCONCH SIZE AND FORM RELATED TO DEVELOPMENTAL MODE

The relationship of features of the protoconch to developmental mode is especially important in prosobranch gastropods. Because the shell grows accretionally, the protoconch is retained throughout the animal's life at the apex of its shell. However, because it is the oldest part of the shell it is often lost by erosion or other environmental damage. That the protoconch at the apex of the adult shell provides important information about embryonic and larval life history has long been recognized. According to the 'shell-apex theory' (Powell 1942; Thorson 1950; recently discussed in detail by Hadfield and Strathmann 1990), 'a high-spired, multiwhorled protoconch, particularly one which exhibits a small initial whorl, a sinusigeral notch, and a change of sculpturing that distinguishes embryonic from larval shell growth, is indicative of a feeding (planktotrophic) veliger stage in the life history. Inflated protoconchs with a large initial whorl and a few whorls without distinctive larval or adult sculpturing are correlated with a non-feeding (lecithotrophic) veliger stage in the life history. While it seems safe to infer a pelagic larval phase from the presence of a multiwhorled sinusigerate protoconch, benthic development cannot be reliably inferred from the presence of a paucispiral inflated protoconch' (Hadfield and Strathmann 1990: p. 252). The latter caveat applies to *Conus*. Both Protoconchs I and II typically lack sculpture and the boundary between them is not visible. Species with larval shells at least 0.75 mm long at hatching, and with as few as 2.5 whorls at metamorphosis have planktotrophic veligers that must feed in order to attain competence to metamorphose (Perron 1981*c*).

Shuto (1974) developed an *a posteriori* model that relates developmental mode in prosobranch gastropods to size and form of the larval shell or protoconch. This model, depicted in graphic form in Fig. 4.11, predicts that species with the ratio of maximum diameter of the protoconch (D) to its number of whorls (V) greater than 1 will be lecithotrophic, and those with D/V less than 0.3 and V greater than 3 will be planktotrophic. Where $0.3 < D/V < 1$, development is predicted to be planktotrophic if $V > 3.5$, and lecithotrophic if $V < 2.25$. The region where V is between 2.25 and 3.5 is bistable; the model allows either developmental mode, reflecting the uncertainty indicated in the quotation from Hadfield and Strathmann (1990) above. Shuto did not explicitly specify developmental modes for values of $D/V < 0.3$ and $V < 3$; presumably this is a second bistable region (Pawlik *et al.* 1988). In Shuto's model, lecithotrophic refers to non-feeding and includes species with completely benthic development as well as those with a short swimming stage.

Data appropriate for testing the fit of *Conus* species to this model are available for only 16

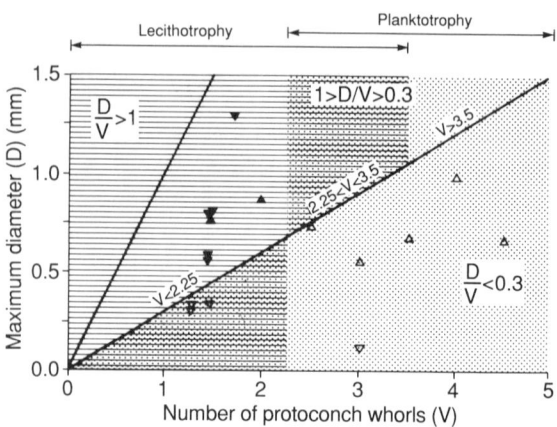

Fig. 4.11 A graphic representation of the model of Shuto (1974) relating the number of protoconch whorls and maximum protoconch diameter to mode of larval development in prosobranch gastropods. Dotted areas are hypothesized regions of planktotrophy. Hatched areas are regions of hypothesized lecithotrophy. Both modes may occur with the combination of conditions indicated by the area where both symbols overlap. Symbols indicate positions of *Conus* species for which both attributes are known. Indo-Pacific species: △, planktonic; ▲, non-planktonic. Species from other regions: ▽, planktonic; ▼, non-planktonic.

species, seven from the IP and nine from Atlantic regions. The positions of these species, superimposed on Fig. 4.11, all conform to the model. Its predictions of developmental mode based on size of the protoconch and number of protoconch whorls will probably hold throughout the genus.

ECOLOGICAL ENERGETICS OF REPRODUCTION

Although our data are inadequate to reveal the fraction of their total energy budget that females of most *Conus* species devote to reproduction, we can estimate some components of reproductive energetics, particularly with reference to female body size.

Conus females clearly continue to grow substantially after becoming sexually mature. Among the 14 species for which we have sufficient data, the shell length of the largest reproductive female observed was 1.2–3.0 times (average 1.8×) that of the smallest reproductive female of the same species (Table 3.2; Figs. 3.2, 3.12; Perron 1982: Fig. 5). Zehra and Perveen (1991) reported length ratios of the largest to smallest reproductive females of both *C. biliosus* and *C. coronatus* of 1.3 (Table 3.2).

The relationship of reproductive effort to female body size has been determined for only four species (*C. abbreviatus*, *C. flavidus*, *C. pennaceus*, and *C. quercinus*; Perron 1982: Fig. 5). The results suggest that reproductive effort, expressed as the ratio of calories in spawn to calories in the body immediately before spawning, changes little with increasing body size.

To determine more generally whether the clutch of eggs produced by a female represents a constant proportion of female body mass, we regressed the logarithms of total egg volume against logarithms of female shell length across species, following a procedure similar to that of Spight and Emlen (1976), who regressed the logarithms of the number of eggs per clutch on the logarithms of female shell length in *Nucella*. Unfortunately, we lacked sufficient data to regress total egg volume directly on female dry weight. Assuming shell shape is constant, a regression slope of 3 indicates investment of a constant fraction of body weight in eggs by females of different sizes and species. For the 47 size-specific fecundity measurements, the observed RMA regression slope was 2.8 (95 per cent confidence limits 2.6, 3.2) (Fig. 4.12). The observed slope indicates that larger females devote similar proportions of their body mass for each egg clutch compared to smaller ones. Clutch volume thus increases with female body size among species, but perhaps at a slightly lower rate than would be predicted from scaling with body mass. In the OLS regression, with a slope of 2.4, female size accounted for 71 per cent of the variance in total volume of eggs produced ($P < 0.0001$).

Most sample sizes were too small to permit within-species comparisons (Appendix Table 1), and only in *C. textile* ($N = 6$) was the OLS regression significant, accounting for 73 per cent of

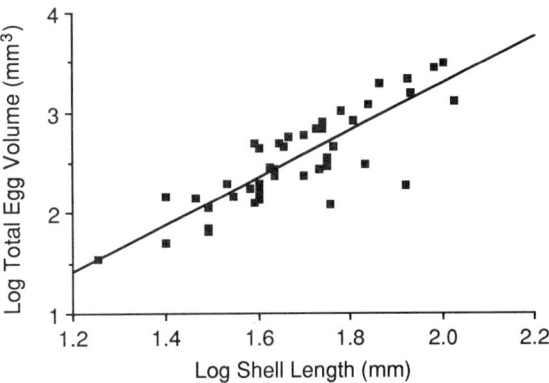

Fig. 4.12 Regression of total egg volume per egg mass on shell length. At the scale of the graph the RMA and OLS regression lines are indistinguishable. RMA: log Y = 2.82 log X − 2.16 (95% CL = 2.40, 3.31). OLS: log Y = 2.38 log X − 1.46 (95% CL = 1.93, 2.84); r^2 = 0.71. $P < 0.0001$). Because of large intraspecific differences between geographically distant populations of some species, the following are plotted separately: *C. eburneus*: Australia, Enewetak; *C. episcopus*: Indonesia, Palau; *C. pennaceus*: Hawaii, India, Indonesia, Maldives. $N = 47$.

the variance in total volume of eggs produced ($F = 10.6$, $P = 0.03$). The observed OLS slope was 3.2; because the sample size was small and the correlation coefficient only 0.85, the RMA slope is 3.7, but 3.0 is within the 95 per cent confidence limits. The proportion of body mass that females of this species devote to reproduction may remain constant or increase over the observed size range of 54–74 mm. This result suggests that if we had the data to permit calculation of size-specific reproductive effort in caloric units, *C. textile* would likely show a pattern similar to the four species noted above as reported by Perron (1982).

TAXONOMIC CONSIDERATIONS

The large genus *Conus* has resisted attempts at division into subgenera by many authors over the past two centuries, and no one has yet ventured to

Table 4.1 Differences in egg diameter and number of eggs per capsule between infrageneric groups in *Conus*, based on Mann–Whitney *U* tests. Lower half matrix gives 2-tailed probabilities of difference in egg diameter (above) and number of eggs per capsule (below). Upper half matrix gives 2-tailed probabilities of differences in adult shell length

Species group	N	Virroconus	Rhizoconus	Lithoconus	C. textile	Pionoconus
Virroconus	11		0.05	0.002	0.002	NS
Rhizoconus	4	NS NS		NS	NS	NS
Lithoconus	4	NS NS	NS NS		NS	0.03
C. textile	11	0.002 0.05	0.002 0.02	0.002 0.02		NS
Pionoconus	5	0.05 0.002	NS 0.01	NS 0.01	NS 0.02	

Small ◀──────────── Egg diameter ────────────▶ Large

generate a phylogenetic hypothesis by cladistic methodology. However, fairly clear groups of closely related species certainly exist. We now examine the interspecific comparisons discussed above in order to determine whether patterns of reproductive biology are congruent within and differ between these species-groups. We apply the infrageneric classification of Marsh (1974); this acknowledges only its accessibility and recency, not its possible validity or superiority over older, alternative schemes.

Our samples include representatives of 20 of the 30 infrageneric taxa Marsh (1974) employs. The largest group (*Virroconus*) contains 11 of the 59 species studied, and *Pionoconus* contains 5, but 15 of the groups are represented by only 1–3 species, making statistical comparisons difficult. Combining the 7 species (comprising 11 separate populations) in three closely related groups (*Regiconus*, *Cylinder*, *Darioconus*) into one (identified here as the '*C. textile* group') gives 5 infrageneric groups each of 4–11 species. Of the 10 possible pairwise comparisons among these, egg diameters of 4 differ significantly (Table 4.1). *Pionoconus* (the *C. magus* group) has the largest eggs, followed in order by the *C. textile* group, *Lithoconus* (the *C. leopardus* group), and *Virroconus* (the *C. ebraeus* group), which includes many species of small adult size. *Rhizoconus* (the *C. miles* group) has the smallest egg diameters.

The number of eggs deposited per capsule varies much more widely than does egg diameter among *Conus* species (Fig. 4.6). Seven of the 10 possible pairwise infrageneric comparisons differ significantly in number of eggs per capsule. These differences could be a corollary of the strong correlation between adult body size and number of eggs per capsule noted in the previous section, but this could explain only one of the significant relationships (*Lithoconus–Pionoconus*; that between *Virroconus* and the *C. textile* group is in the opposite direction; Table 4.1).

5 RELATIONSHIPS OF DEVELOPMENT AND BIOGEOGRAPHIC PATTERNS

The Indo-Pacific marine biogeographic realm extends halfway around the world and across 60° of latitude, from the Pitcairn Islands and Hawaii, across the Pacific and Indian Ocean basins, to East Africa and the Red Sea. It covers about 10^8 km^2, or more than a quarter of the earth's total ocean area, yet its marine biota is remarkably homogeneous over this vast extent. As we noted in Chapter 1, some species of *Conus* appear to maintain gene flow over the entire IP region, while others are more narrowly distributed, some to single archipelagoes, basins such as the Sulu Sea, or narrow stretches of continental coastline.

In this chapter, we examine the biogeographic patterns of IP *Conus* in light of the information on development and dispersal capability discussed above. We address the broader question, how can we explain present distribution patterns of tropical IP marine benthic invertebrates? That is, what factors and processes determine the components of species distribution patterns, notably the sizes, shapes, and extent of their geographic ranges? In particular, in light of the data presented in the preceding chapters, are the geographic ranges of species related to their developmental modes?

Hypotheses that seek to explain observed differences among the geographic distributions of organisms may be characterized as either historical or ecological. This distinction is mainly one of temporal scale. Alternatively, distribution patterns may be hypothesized as being determined either by dispersal, i.e. by the organisms' ability to extend their range, or by vicariance, i.e. by disruption or separation of a formerly continuous range caused by environmental factors external to the organisms themselves. Both types of hypotheses recognize or assume that distributions are also constrained by ecological determinants—the range occupied must satisfy the organisms' ecological requirements. Conversely, ecologically mediated selective factors can act only on species whose propagules can reach the environment—the pool of available colonist species. Ecological attributes that could constrain the geographic range of marine gastropods include food habits: predatory species may be distributed more widely than species at lower trophic levels; substratum: sand-bottom species may be distributed more widely than rock dwellers; depth: low intertidal species may be distributed more widely than those higher on shore, and shallow subtidal species, more widely than deeper dwellers; anti-predator defences: species with thick shells, narrow apertures, and other such devices may be distributed more widely than those lacking them (Vermeij 1978, 1987).

In explaining the distribution patterns of closely related species, the vicariance hypothesis requires dispersal in the earlier history of the taxon, a long time before the imposition of the barriers that lead to speciation events, while the dispersal hypothesis requires interruption of gene flow soon after a dispersal event for speciation to occur. Temporal scale is an important element of this distinction. The dispersal hypothesis is appropriate to shorter ('ecological') and the vicariance hypothesis to longer ('evolutionary') time-frames. Dispersal events occur within the lifetime of an individual, while extrinsic, vicariance events typically involve long-term alterations of the face of the earth.

Strong evidence exists in support of both dispersal and vicariance hypotheses in different taxa, and sometimes in the same taxon (Futuyma 1986; Stace 1989), so the two kinds of hypotheses are not mutually exclusive. They are also notoriously hard to refute.

We are convinced that dispersal necessarily plays an important role in determining the distribution patterns of marine invertebrates that are benthic as adults, have a planktonic larval stage, and are restricted to the shallow waters around isolated volcanic oceanic islands that rise from great depths of the sea. Most of the species of *Conus* discussed in Chapter 3 share all of these attributes. Benthic and rather sedentary as adults, these species' distribution patterns must depend on the dispersal larval phase of the life history for recruitment and maintenance of populations.

We thus apply the data developed on dispersal ability to test a prediction of the dispersal hypothesis: the extent of a species' geographic range should be proportional to its dispersal ability. For benthic marine invertebrates, dispersal ability may be assumed to be proportional to the duration of the planktonic larval phase. A positive relationship between dispersal ability and extent of geographic range, for example in comparisons across species of a species-rich taxon, supports the hypothesis that dispersal ability is an important determinant of distribution pattern. The hypothesis is refutable: if species with long planktonic stages are confined to narrow geographic regions, vicariance or ecological determinism must exert a more important influence on geographic range than does dispersal ability. Thus we now pose the question, do species of *Conus* whose small eggs and early, altricial hatching commit them to a long pelagic period disperse more widely and occupy broader ranges than more precocial species, with short pelagic stages, or none at all in their life histories. We know nothing of the maximum or realized (Scheltema 1986*a*) durations of planktonic larval stages in *Conus*, and there is at present no way to obtain such knowledge. Theoretical treatments (Jackson and Strathmann 1981) suggest that the length of the competent stage should be roughly proportional to that of the precompetent stage, which we can measure or estimate. We discuss below the experimental studies of Pechenik (e.g. 1980) on prolonging the competent phase in other gastropods.

Unfortunately the other necessary variable, the geographic distribution, is also rather poorly known for most *Conus* species. The most detailed published lists of locality records are those of Dautzenberg (1937), but these include many obsolete, erroneous data. Coomans *et al.* (1979–1986) published range maps of about 80 IP species based on the literature and specimens in the collection of the Zoological Museum of the University of Amsterdam. In the preparation of a taxonomic treatment of IP *Conus* (Röckel *et al.* 1993), we compiled and critically reviewed records from the literature and from recent collections in several major museums. We transferred these data to maps of the IP region and plotted range maps, using the basic criterion of minimum convex polygons or hulls (following Sedgewick 1983), but we subtracted large land areas that protruded into the hulls. Figure 5.2 indicates that for *C. coronatus*, Arabia, India, Indo-China, and parts of East Africa, Australia, and China were treated in this way. The resulting range maps were then digitized and their areas computed in arbitrary units. The areas were determined somewhat crudely, because the Mercator projections used as base maps account minimally for the effect of the earth's curvature on map area. Area at higher latitudes is thus overestimated, but because most of the species considered have exclusively tropical ranges, and all extratropical parts of ranges of the species considered are quite small, this error is minimized.

Habitats suitable for benthic adults of species with narrow bathymetric ranges of course comprise but a small fraction of the largely oceanic areas mapped. The fact that suitable adult habitats are very discontinuously distributed causes 'the problem of range interpolation' (Rosen 1988). Our methodology solves this by circumscribing all adult locality records, so that the range maps of benthic adults model the hypothesis that the species maintains gene flow and dispersal ability throughout the mapped range.

PATTERNS OF GEOGRAPHIC RANGES OF INDO-PACIFIC *CONUS*

Based on data derived from digitized map areas, Fig. 5.1 compares the absolute geographic ranges of IP *Conus*, i.e. without regard to specific location. The normalized abscissa expresses range as the proportion of the area of the most widely distributed species, *C. ebraeus*, occupied by each other

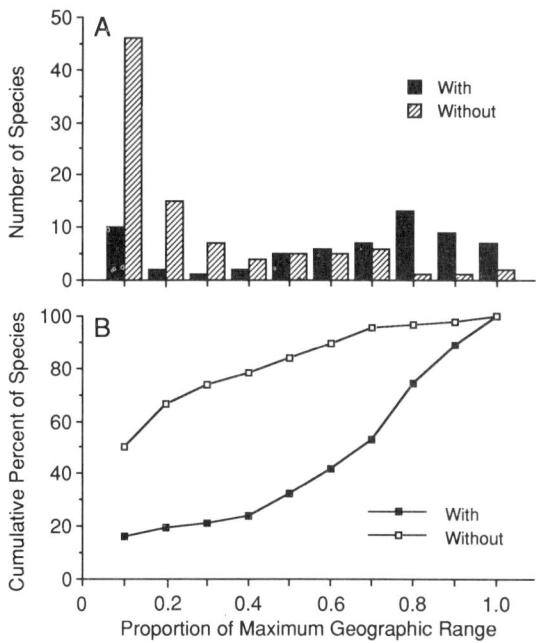

Fig. 5.1 Frequency distributions (A) and cumulative curves (B) of the areal extent of geographic ranges of Indo-Pacific *Conus* species. Dashed histograms and curve with open squares: all IP species lacking data on reproductive biology that have been mapped ($N = 92$). Solid histograms and curve with solid squares: species with data on reproductive biology discussed here and listed in Appendix Table 1. ($N = 62$).

species. (The portion of the range of *C. ebraeus* outside the IP, i.e. east of the East Pacific Barrier, is omitted.)

Of the 154 species for which we have reasonable data on geographic distribution, 52 per cent have ranges that are 20 per cent or less of the maximum area (Fig. 5.1A). Many taxa of marine as well as terrestrial animals have similar but even more skewed frequency distributions of range size, with most species narrowly distributed and rather few occupying broad geographic areas (Gaston 1990).

The solid histograms in Fig. 5.1A show the frequency distribution, and the lower curve in Fig. 5.1B shows a cumulative distribution, of range sizes for the 62 species for which some data on

reproduction and development are available (see Appendix Table 1). The graph indicates a definite bias—a higher proportion of species with reproductive data range more widely in the IP region than do species for which we lack information on reproduction (hatched histograms in Fig. 5.1A, upper curve in Fig. 5.1B: K-S test: $P < 0.001$). This is not unexpected; a widely distributed species is more likely to be observed somewhere in its range than a narrowly restricted species. However, the sample of species with reproductive data includes species that encompass the entire spectrum of range sizes.

The most widely distributed species in the genus are *C. ebraeus* and *C. coronatus*. Both occupy virtually the entire IP region (Figs. 5.2A,B). *Conus ebraeus* occurs at Pitcairn Island, one empty shell has been collected at Easter Island, and it maintains breeding populations on the west coast of Central America and in the eastern Pacific islands. In general, *Conus* species distributed widely over the tropical Pacific islands tend to occur throughout much of the Indian Ocean as well. Figures 5.2C and D show the ranges of two species with planktonic larvae whose ranges are about 75 per cent and 50 per cent of that occupied by *C. ebraeus*. They also occupy widely separated archipelagoes but do not extend quite as far eastward or westward. Some species have more restricted ranges, occurring on the continental coasts and continental islands of the eastern Eurasian Plate, e.g. *C. cinereus* (Fig. 5.2E), which likely lacks a planktonic larva. A few are endemic to oceanic archipelagoes e.g. *C. abbreviatus* (Fig. 5.2E), and a few are restricted to more continuous or linear continental coastal environments e.g. *C. biliosus* (Fig. 5.2F). Despite their narrow ranges, the latter two species have planktonic larvae.

DISPERSAL AND DISTRIBUTION: TESTING THE DISPERSAL HYPOTHESIS

In order to test the hypothesis that geographic range is correlated with dispersal ability, we now examine the relationship of geographic distribution patterns, determined from the records outlined above, to

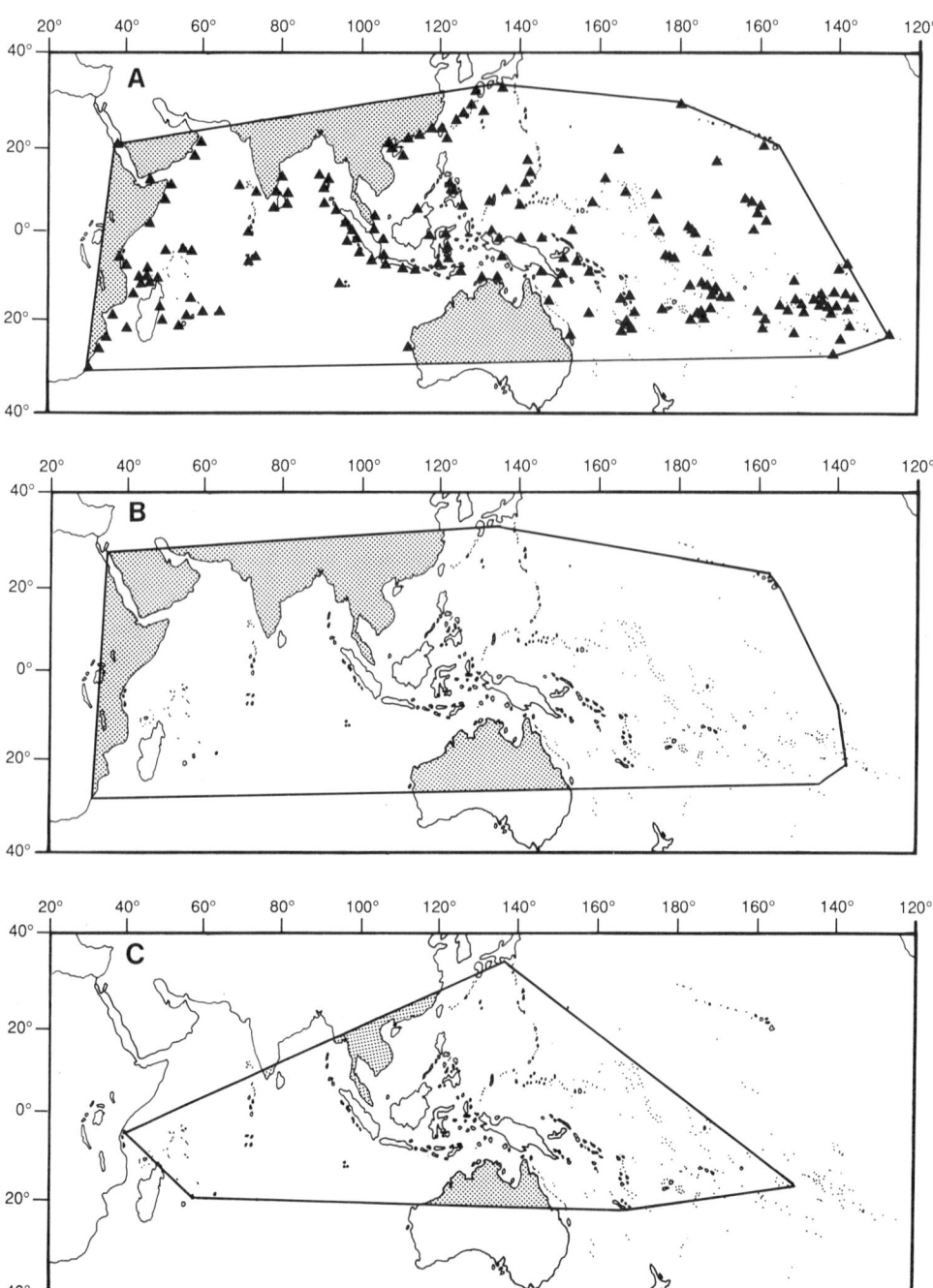

Fig. 5.2 Examples of geographic ranges of IP *Conus* species with different distribution patterns. Areas of geographic ranges were computed from the convex hulls shown. Figures in parentheses indicate the fraction of the area of the maximum (*C. ebraeus*) range occupied by each species. (A,B) Species with the maximum ranges: *C. ebraeus* (A: 1.00) and *C. coronatus* (B: 0.95). (C,D) Widely distributed species: *C. litteratus* (C: 0.72) and

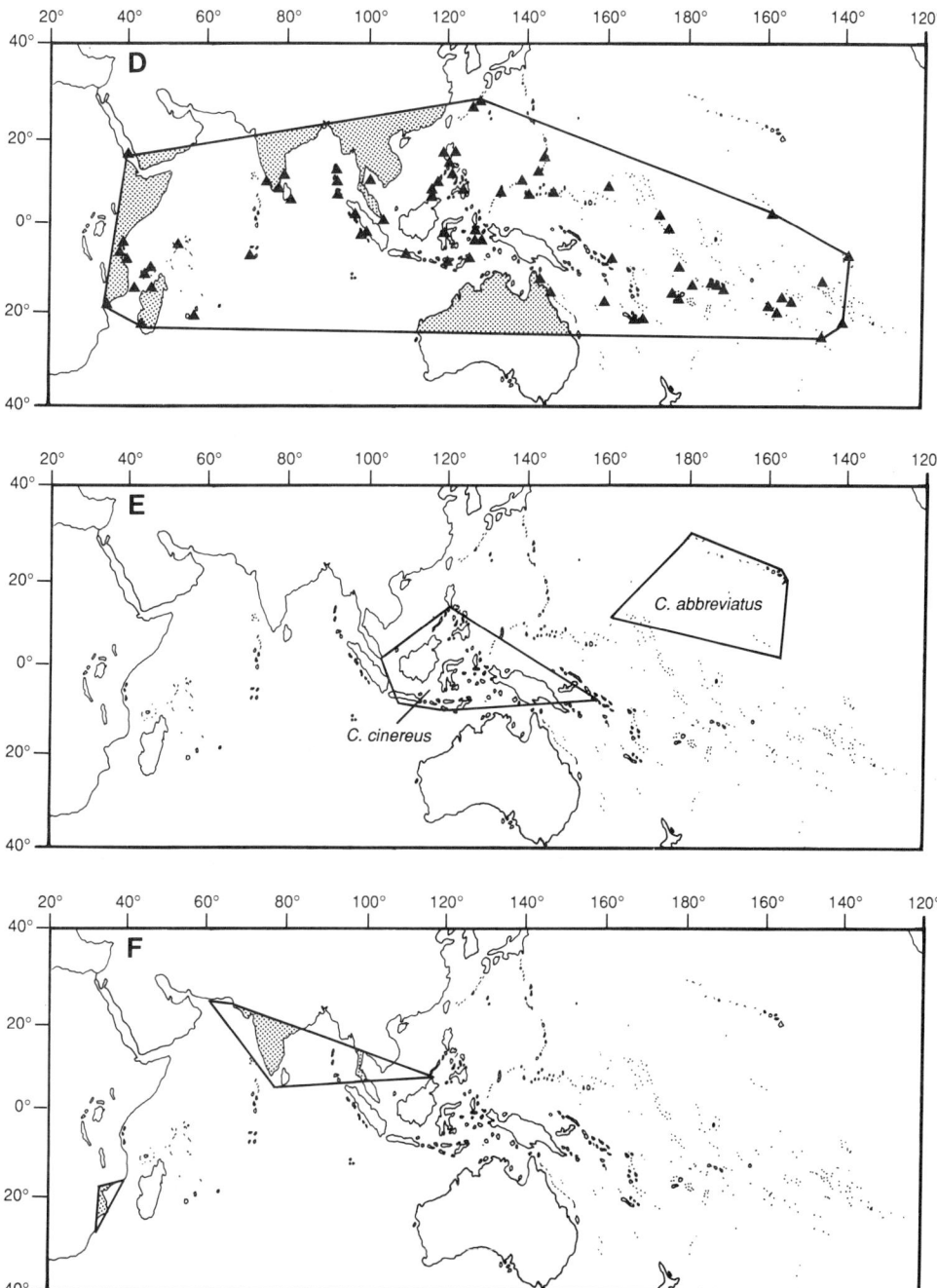

C. balteatus (D: 0.53). (E) Species restricted to oceanic archipelagoes: *C. abbreviatus* (0.07), and to continental coasts and islands of the eastern Eurasian plate: *C. cinereus* (0.08). (F) Species restricted primarily to linear continental coasts: *C. biliosus* (0.04). The range map of *C. ebraeus* omits eastern Pacific records of this trans-Pacific species.

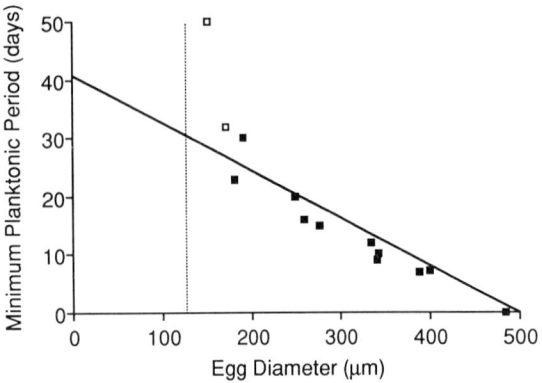

Fig. 5.3 Relationship between minimum planktonic period and egg diameter in *Conus* species from Hawaii (H) and Palau (P). Right to left: *C. pennaceus* (H); *C. marmoreus* (P); *C. consors* (P); *C. bandanus* (H); *C. glans* (P); *C. omaria* (P); *C. episcopus* (P); *C. textile* (H); *C. striatus* (H); *C. flavidus* (H); *C. quercinus* (H); *C. abbreviatus* (H); *C. lividus* (H). The last two species are omitted from the calculated regression (see text). The vertical dashed line indicates the minimum observed egg diameter in *Conus* (Perron and Kohn 1985).

dispersal ability, determined from the observations and estimates of the precompetent planktonic periods (T_P) described in Chapter 4. Specifically, we use the regression of T_P on egg diameter for the 13 species reared through the entire precompetent period to metamorphosis (Fig. 5.3) as a model to estimate T_P in other species of the genus, for which we have data on egg size and some aspects of embryonic and larval development, but for which we do not know T_P from laboratory observations. The total developmental trajectories from oviposition to hatching to settlement are illustrated in Fig. 4.7, which also distinguishes between entries based entirely on direct observation and those that estimate T_P.

BIOGEOGRAPHIC PATTERNS AS TESTS

Several biogeographic patterns of *Conus* species in the IP region are amenable to testing the dispersal hypothesis in this way. We examined the following three:

(1) species with a predominantly linear or one-dimensional distribution, i.e. narrowly distributed along continuous continental shoreline, as opposed to species with a predominantly two-dimensional distribution, i.e. widely distributed among oceanic islands;

(2) species confined to the predominantly continental shorelines plus continental island groups of the eastern margin of the Eurasian lithospheric plate, as opposed to species that extend eastward to island groups on the Philippine and Pacific plates;

(3) species that simply vary in area of range occupied, without regard to specific location within the IP.

ONE-DIMENSIONAL VS. TWO-DIMENSIONAL DISTRIBUTION PATTERNS

The dispersal hypothesis predicts that:

(1) species with two-dimensional ranges in oceanic islands will have small eggs and hence longer precompetent planktonic durations than those with more linear ranges along continental shorelines;

(2) non-planktonic development will be more characteristic of species in the latter than in the former environment.

The data, organized geographically in four oceanic and three continental regions (Table 5.1) are consistent with these predictions. Species distributed among the oceanic islands have an average T_P of 2–4 weeks, and only one of the 41 species (some species appear in more than one column of Table 5.1A) has non-planktonic development. Of the 36 species representing coastlines of continents and continental islands (again some species appear in more than one column) in Table 5.1B, 20 per cent lack a planktonic larval stage. To test the significance of this difference, we lumped the data for the

Table 5.1 Egg diameters, precompetent planktonic periods, and incidence of non-planktonic development in *Conus* from different Indo-Pacific areas (means ± SD)

A. Oceanic areas	Indian Ocean	Palau	Marshall Islands Ponape	Hawaii
$\bar{X}$ egg diameter (μm)	187 ± 65	241 ± 75	281 ± 119	206 ± 87
$\bar{X}$ planktonic period (days)	24 ± 5	19 ± 7	16 ± 10	24 ± 10
No. of species with				
non-planktonic development	0	0	0	1
planktonic development	15	22	10	18

B. Continental areas	India, Sri Lanka, Pakistan, Persian Gulf	Indonesia, Thailand	Philippines
Species with planktonic larvae:			
$\bar{X}$ egg diameter (μm)	205 ± 95	246 ± 85	243 ± 82
$\bar{X}$ planktonic period (days)	23 ± 8	19 ± 7	20 ± 7
No. of species	9	15	13
Species without planktonic larvae:			
$\bar{X}$ egg diameter (μm)	569 ± 75	563	561 ± 64
No. of species	3	1	6

different areas within each part of Table 5.1 but entered each species only once in each part. The G test for independence with Williams correction (Sokal and Rohlf 1981) indicates that the frequencies of planktonic and non-planktonic development in oceanic vs. continental areas differ highly significantly ($G = 6.05$; $P < 0.025$).

Among the species with planktonic larvae, T_P of species inhabiting continental shores is slightly shorter (average of 20 days) than among the oceanic island species (average of 22 days), but not significantly so (Mann–Whitney U test: $P = 0.16$).

The one widely distributed species with non-planktonic development indicated in Table 5.1A (*C. pennaceus* from Hawaii) has, like most molluscivorous species, a Type II egg mass. Since only a few capsules are directly affixed to the substratum in Type II egg masses, these structures are relatively easily dislodged and set adrift. If some of the capsules in a Type II egg mass become punctured and filled with air, the entire mass is capable of floating for long periods of time. Such drifting egg masses have been documented (Kohn 1961a) and we have observed them on several subsequent occasions. In species of *Conus* lacking planktonic larvae, such egg masses could provide an alternate means of dispersal. Egg mass floating seems a likely mechanism for the dispersal of this species to the isolated Hawaiian Islands. Since *C. pennaceus* has an intracapsular developmental period of up to 26 days (Appendix Table 1), and since posthatching juveniles can survive for at least another 26 days without food (Perron 1981b), a drifting egg mass could carry living colonists for more than 50 days.

We address the question of whether rafting is generally an important factor in species distributions more fully in a subsequent section.

Studies addressing the dimensionality of distribution of species with and without planktonic larvae in other taxa and regions are rare, but Scheltema (1986b, 1989) showed that prosobranch gastropod species of North Carolina with non-planktonic development are mainly restricted to the US Atlantic coastline, while species with planktonic larvae are more widely distributed. A high proportion of species (about 80 per cent) with long-term planktonic veligers (> 6 weeks) have amphi-Atlantic distributions, that is they occur along both the American and African Atlantic coasts.

To summarize, the restricted distribution patterns of most IP *Conus* species having non-planktonic development are adequately explained in terms of their limited dispersal potential, although a few exceptions are discussed at the end of this chapter. Likewise the very broad distributions of species occupying most or virtually all of the vast tropical Indian and Pacific Ocean basins are consistent with the predominance of an obligate feeding planktonic larval stage in their life histories.

RELATIONSHIP OF PATTERNS TO LITHOSPHERIC PLATES

A strictly dispersalist interpretation of *Conus* biogeography would require that no IP species with a lengthy planktonic larval stage should be absent from the islands of the Pacific plate (Perron and Kohn 1985). Lithospheric plate boundaries have been invoked as important factors in the distribution patterns of other animal groups that share the same habitats as *Conus*, particularly coral reef fishes (Springer 1982) and also certain gastropod groups (Kay 1990).

To examine the relationship of *Conus* distribution patterns and larval development modes to lithospheric plate boundaries we select as examples the *Conus* fauna of two archipelagoes on different plates. These are the Philippine Islands, on the eastern margin of the Eurasian plate (not on the Philippine plate) and New Caledonia and nearby islands, on the eastern margin of the Indo-Australian plate. We ask whether developmental modes differ in those species that do vs. those that do not extend their ranges to the islands on adjacent plates.

The Philippines *Conus* fauna probably comprises about 160 species. We have some data on geographic and bathymetric ranges of 135 of these, of which 100 occur in shallow water, that is in depths less than about 20 m (Springsteen and Leobrera 1986; Walls [1979]). Of the latter, 69 extend their ranges to islands of at least three of the four other major lithospheric plates that underlie the IP region (the Philippine, Pacific, Indo-Australian and African plates). Six shallow-water species do not reach even Palau or Guam on the eastern margin of the Pacific plate (Table 5.2). The remaining 25 species occur only on the Indo-Australian (11) or on two additional plates (14).

Of the 69 shallow-water Philippine species that also occur on islands of four or all five plates, developmental mode is known for 47. All of these except *C. magus* (98 per cent) have planktonic larvae (Table 5.2). In striking contrast, both of the shallow water species whose ranges are confined to the Eurasian plate and for which egg size is known have non-planktonic development. Despite the very small sample size of the latter group, the difference is significant (G test of independence with Williams correction; $G = 5.5$; $P < 0.025$). Thus almost all of the shallow water *Conus* species in the Philippines whose developmental mode is known, either extend their ranges onto 3–4 other plates and have planktonic development, or are restricted to the Eurasian plate and have non-planktonic development.

The second example, New Caledonia, on the eastern margin of the Indo-Australian plate, is somewhat more complex. Estival (1981) lists a total of 93 species, of which 85 occur at depths less than 20 m. As in the Philippines, nearly all species (95 per cent) extend to archipelagoes on one or more adjacent plates, and about 80 per cent (68 species) occur on four or all five IP plates. The developmental mode of 48 of these species is

Table 5.2 Distribution of *Conus* species with respect to lithospheric plates: the Philippine Islands and New Caledonia as examples

	Approximate no. of species			
The genus *Conus* (extant species)	500			
Indo-Pacific Region	250			
Philippines: Total	160	With planktonic larvae	Without planktonic larvae	Development unknown
Shallow water:	100			
Occurring on 4 or 5 lithospheric plates	69	46	1	(22)
Occurring on 2 or 3 lithospheric plates	25	1	3	(21)
Confined to Eurasian lithospheric plate	6	0	2	(4)
	100			
New Caledonia: Total	93	With planktonic larvae	Without planktonic larvae	Development unknown
Shallow water:	85			
Occurring on 4 or 5 lithospheric plates	68	48	1	(19)
Occurring on 2 or 3 lithospheric plates	13	0	2	(11)
Confined to Indo-Australian lithospheric plate	4	0	0	(4)
	85			

known; all but one (*C. magus*) have planktonic larvae (Table 5.2). Thirteen species are confined to the Indo-Australian plate or occur on one or two other plates, but unfortunately developmental mode is known only for two of them (Table 5.2), *C. achatinus* and *C. cinereus*, both of which have non-planktonic development. However, almost all of the New Caledonian species that are widely distributed to islands on three or four other plates and whose development is known have planktonic larvae.

RELATIONSHIP OF ABSOLUTE RANGE AREA TO DEVELOPMENTAL MODE

In this section we examine the simplest prediction of the dispersal hypothesis, that the longer precompetent planktonic larval period a species has, the broader should be its geographic range. Here we consider only the absolute area occupied by IP species, without regard to position of the ranges on the globe. The dependent variable is range area, determined from the maps digitized for Figs. 5.1

and 5.2, and normalized to the area occupied by the most widely distributed species, *C. ebraeus*. The independent variable is the minimum precompetent period, either determined in the laboratory (Fig. 5.3) or estimated from egg diameter (Figs. 4.1, 4.7). The OLS regression for the entire data set of 61 species (Fig. 5.4) is highly significant and explains 39 per cent of the variation in normalized range area ($P = 0.0001$; Table 5.3). If the proportions of maximum area are arcsin transformed and the durations of precompent period are log transformed to normalize the variances, the regression accounts for 43 per cent of the variance in geographic range area and is also significant at the 0.0001 level (Table 5.3).

Floating and rafting may facilitate dispersal in a number of marine invertebrate groups with brief or no planktonic stages in the life history (Highsmith 1985). The hypothesis that rafting of egg masses enhances distribution of *Conus* leads to the prediction that species with Types II and III egg masses should have broader geographic ranges than would be predicted by egg size alone. Or, for a given egg size, species with Types II and III egg masses should be more broadly distributed than species with Type I egg masses. In Fig. 5.4, different symbols identify species according to type of egg mass. When geographic range is regressed on precompetent duration separately for these groups, there is no significant difference in the slopes or intercepts. This is true whether or not the data are transformed to the arcsin of the proportion of maximum geographic range and the log of minimum planktonic period (Table 5.3). Thus statistically, species with egg masses of the types that are most likely to be transported by rafting do not have broader ranges than those in which each capsule is firmly affixed to the substratum.

To summarize, the relationship between area of range and length of the precompetent stage in IP *Conus* is positive and significant. Species with greater potential for larval dispersal thus have significantly broader geographic ranges. The most striking exceptions, or outliers in Fig. 5.4, are the

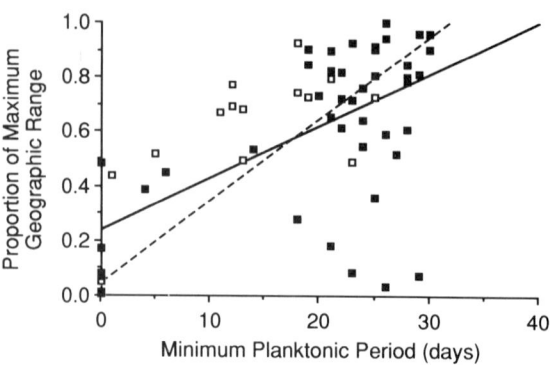

Fig. 5.4 Relationship of geographic range area (normalized to the maximum range, that of *C. ebraeus*) to minimum planktonic or precompetent period in Indo-Pacific *Conus*. $r = 0.62$; $N = 61$. Open symbols: species with Types II and III egg masses, which are susceptible to rafting. Solid symbols: species with Type I egg masses, which are extremely unlikely to be rafted. For regression statistics based on these and transformed data, and analysis of covariance, see Table 5.3.

three species at the lower right, which have small eggs, and long observed or estimated planktonic stages, but narrow geographic ranges. We have only observed one of these directly, *C. abbreviatus*, with a precompetent period of 26–32 days. It is confined mainly to the Hawaiian archipelago but has been collected at Enewetak in the Marshall Islands. The eggs of the narrowly distributed *C. biliosus* (Fig. 5.2F) are 172 μm in diameter according to Zehra and Perveen (1991), giving an estimated precompetent period of 26 days. The eggs of *C. suratensis*, in an egg mass collected and identified by J. E. Norton, were 207 μm in diameter, giving an estimated precompetent period of 23 days. This species is known from the Philippines, Malay peninsula, Sri Lanka, and Madagascar. In some species, such as *C. pennaceus*, the accidental dislodgement and rafting of egg masses may provide an alternative but infrequent mode of long-distance dispersal, but this has probably not affected the distribution of very many species.

Table 5.3 Relationships between geographic range area and minimum planktonic duration of Indo-Pacific *Conus* species. Regressions of Y = geographic range area normalized to that of *C. ebraeus*, on X = precompetent planktonic period in days. In regressions of transformed data, $Y' = [\sin \theta]^2$, where $\theta = \arcsin(\sqrt{Y})$, and $X' = \log(X + 1)$

Treatment	No. of species		Regression	r^2	F	P	95% CL
All species	61	OLS	$Y = 0.02X + 0.24$	0.39	37.5	0.0001	0.01, 0.03
		RMA	$Y = 0.03X + 0.05$	—	—	—	0.02, 0.04
		OLS	$Y' = 0.51X' + 0.11$	0.43	44.5	0.0001	0.36, 0.67
		RMA	$Y' = 0.78X' - 0.18$	—	—	—	0.64, 0.95
Species with Type I egg capsules	44	OLS	$Y = 0.02X + 0.23$	0.32	19.7	0.0001	0.01, 0.03
		RMA	$Y = 0.03X - 0.08$	—	—	—	0.03, 0.04
		OLS	$Y' = 0.48X' + 0.14$	0.30	18.3	0.0001	0.26, 0.71
		RMA	$Y' = 0.88X' - 0.34$	—	—	—	0.74, 1.19
Species with Types II and III egg capsules	17	OLS	$Y = 0.03X + 0.20$	0.65	27.5	0.0001	0.02, 0.04
		RMA	$Y = 0.03X + 0.12$	—	—	—	0.03, 0.05
		OLS	$Y' = 0.59X' + 0.07$	0.80	61.2	0.0001	0.43, 0.75
		RMA	$Y' = 0.66X' + 0.10$	—	—	—	0.51, 0.83

Analysis of covariance		Regression	Slopes		Intercepts	
			F	P	F	P
Type I vs. Types II and III		Simple linear	0.05	0.8	1.80	0.2
		Transformed	0.13	0.7	0.13	0.7

6 *CONUS* DEVELOPMENT OUTSIDE THE INDO-PACIFIC REGION

Relatively little is known of the reproductive biology of *Conus* outside the IP region, even though the number of species occurring elsewhere in the world probably approximately equals the number of IP species.

THE WESTERN ATLANTIC

Lebour (1945) and Lewis (1960) described the egg capsules and larvae of *Conus mus* from Bermuda and Barbados respectively, D'Asaro (1970) reported briefly on two Florida species, and studies by Bandel (1975*b*, 1976), Penchaszadeh (1984) and Vink and Cosel (1985) provide limited information on nine additional species from the Caribbean (Appendix Table 2). These reports provide measurements of egg size for only three species, but in some cases it can be estimated from protoconch size, using the regression based on IP species (Fig. 4.8) as a model. Four of the western Atlantic species, *C. centurio*, *C. ermineus*, *C. mus*, and *C. regius*, hatch as free-swimming veliger larvae. Their estimated egg diameters are 203–242 μm. Lebour (1945) stated that the egg of *C. mus* 'measures *c*. 0.1 mm across' and Lewis (1960) gave the diameter as 0.12 mm. The maximum size of the veliger shells, given as 0.24 mm by Lewis, 0.32 mm by Lebour and 0.34 mm by Bandel (1975*b*) would indicate an egg diameter range of 155–240 μm according to the IP regression (Fig. 4.8). The remaining six species have (*C. puncticulatus*, *C. spurius*) or probably have (*C. aurantius*, *C. cedonulli*, *C. mappa*, *C. pseudaurantius*) non-planktonic development. Their estimated egg sizes are about 550–600 μm, but the data from *C. mus* suggest that these may be overestimates.

THE EASTERN ATLANTIC AND MEDITERRANEAN

Until a very few years ago data on the reproduction of eastern Atlantic and Mediterranean *Conus* were quite sparse. A recent surge of research, especially by Rolán (1985, 1986*a*,*b*, 1990) has increased the number of species with some information to 23. However, many of these are described as new species with unusually narrow geographic ranges within the Cape Verde Islands, and confirmation of their taxonomic status will depend on the results of future biological studies.

Rolán's results indicate that all of the Cape Verde Islands species, as well as *C. guinaicus* from the Canary Islands and the Mediterranean *C. ventricosus*, share a common pattern of small clutches of large eggs and a demonstrated or probable lack of a planktonic stage (Appendix Table 2). All species from the Cape Verde Islands studied by Rolán (1985–90) develop to the veliconcha stage or beyond within the egg capsule, and at least three are known to lack a planktonic larva. The only known exception in the region is the amphi-Atlantic *C. ermineus* (Rolán 1985, 1986*b*).

Rolán (1985) described the development of only *C. cuneolus* in detail, and it is unlike any known IP species in several respects. Although egg size was not reported, the embryonic period is nearly twice as long as that of the longest known IP species (45 days vs. a maximum of 26 for *C. pennaceus*; Fig. 4.7, Appendix Table 1), and development proceeds beyond the veliconcha stage to extremely well-developed juveniles within the egg capsule. Prior to hatching, the embryos crawl over the inner walls of the capsule with the large, well-developed foot. The estimated egg diameter of four of the seven Cape Verde species is 347–477 μm. The IP model (Fig. 4.1) would predict them to have planktonic larvae, but it should be noted that the size of the adults (maximum of 19–23 mm) and of their egg capsules (2–6 mm high) and the number of eggs per capsule (4–12) (Appendix Table 2) are all unusually small for the genus. As is the case in other marine invertebrate groups, members of the smallest species may tend to produce few, large eggs and have non-planktonic development.

SOUTHERN AUSTRALIA

Wilson and Gillett (1971) illustrated the spawn and newly hatched benthic juveniles of the temperate Australian species *Conus papilliferus*. The shells of the latter appear to be about 1.3 × 0.7 mm. Huish (1978) reported the egg capsules of *C. papilliferus* collected at Minnie Water, New South Wales in August 1978, to contain 10 eggs, each 1000 µm in diameter. Huish (1978) also collected egg capsules of the IP species *C. coronatus* at the same site in June 1978. Their eggs were the same size as in tropical populations of the same species (Appendix Table 1).

Wilson and Gillett (1971) stated that *Conus anemone* also had non-planktonic development, and Huish (1978) briefly noted egg diameter (590 µm) and number per capsule (65) in *C. anemone* from Portsea, Victoria.

Kohn (1993) studied reproductive and developmental patterns of three species in temperate Western Australia, *C. anemone*, *C. dorreensis* and *C. klemae*. All deposit Type I egg masses on the undersides of flat rocks on intertidal shore platforms, and *C. dorreensis* also oviposits in clumps of algal turf. Of the three species, *C. dorreensis* has the smallest eggs (152 µm in diameter); the average number per egg mass was 40 000, with each capsule containing 1000–2000 eggs. The prehatching duration is 12–17 days, followed by hatching of small (250 µm) planktotrophic veligers. The eggs of *C. klemae* are 280 µm in diameter; one egg mass contained about 61 000 eggs, with about 1100 in each capsule. These hatch as well-developed planktonic veligers with shallowly bilobed velar lobes and shells 450 µm in maximum diameter.

The eggs of *Conus anemone* from Western Australia are ovoid, averaging 790 × 710 µm, larger than those reported by Huish (1978) from Victoria. The elongate egg capsules of *C. anemone* (illustrated by Smith *et al.* 1989) are thick and tough and contained an average of only 19 eggs. The total number per egg mass is estimated at only about 300. The eggs hatch after an unknown prehatching developmental period as veliconchas with shells 1.6 mm long. Each lobe of the broad velum is divided by a midlateral constriction. The large foot has a distinct propodium and metapodium and an obvious anterior pedal mucus gland. The veliconchas are able to crawl but apparently unable to swim immediately upon hatching. Kohn (1993) gives more detailed information on the timing of developmental events and illustrates the veliconcha and juvenile shells of *C. anemone*.

SOUTHERN AFRICA

Kilburn (1971) described the egg capsules and veliconcha shells of three high latitude (about 35° S) species from the Cape of Good Hope. All have large eggs and almost certainly non-planktonic development. The egg capsules of *C. scitulus* Reeve are rather typical for the genus, but they contain only 1–5 embryos; their veliconcha shells are 2.6 × 1.5 mm. The eggs of *C. tinianus* Hwass in Bruguière are the largest known for the genus, averaging 1 mm in diameter. The egg capsules of *C. mozambicus* Hwass in Bruguière are unusual in form: 'Each is oblong-ovate in shape, with one side markedly more convex than the other, tapering apically to a narrowly rounded summit, and basally to a thick, laterally flattened stalk; the whole capsule inclines at an angle to the basal disc, which does not appear to join adjacent capsules together' (Kilburn 1971). Kilburn and Rippey (1982) stated that this species lacks a planktonic larva.

Table 6.1 Developmental modes of *Conus* in different geographic areas: comparisons of numbers of species with planktonic and non-planktonic development. Data from Appendix Tables 1 and 2

Region	Number of species	
	With planktonic larva	With non-planktonic development
Indo-Pacific	51	11
Western Atlantic	4	6
Eastern Atlantic	1	4
Temperate Australia	2	2
Southern Africa	0	2

THE EASTERN PACIFIC

No species of *Conus* is circumtropical, but a few IP species occur, and probably maintain breeding populations, in this region, both on tropical eastern Pacific islands and along the Pacific coast of Central America (Emerson 1978, 1983). While reproduction and development of these eastern Pacific populations remain completely unknown—as is the case for the some 26 species restricted to the Panamic zoogeographic province (Keen 1971)—we may ask whether IP populations of species that also occur in the eastern Pacific have larvae likely to be dispersed widely.

All three IP species with eastern Pacific range extensions (*C. chaldaeus*, *C. ebraeus*, and *C. tessulatus*) occur both on offshore islands (Clipperton, Galapagos) and along the coast of Central America (Emerson 1978, 1983). The estimated durations of the precompetent stages of these species in the IP (Appendix Table 1), are 25, 26, and 19 days, respectively. The first two exceed the median of all IP *Conus* of 22 days, but their durations are no longer than many other species that have not crossed the 'east Pacific Barrier' (Fig. 4.1).

To our knowledge, observations on oviposition of only one eastern Pacific species, *C. purpurascens* have been reported (Nybakken 1970). It resembles its likely closest relative, the western Atlantic *C. ermineus*, in having several thousand small eggs per capsule that hatch as small veligers.

GEOGRAPHIC PROVINCE, LATITUDE, AND DEVELOPMENTAL MODE

Outside the IP region, *Conus* includes some temperate as well as tropical species, allowing examination of developmental modes over a broader latitudinal range. Unfortunately, egg sizes of only 13 species outside the IP region are known (Appendix Table 2); they are not significantly correlated with latitude ($r_s = 0.16$). Size at hatching is known for far more species (30, with 34 records; Appendix Table 2) from these regions (eastern Atlantic and Mediterranean, western Atlantic, South Africa, and temperate Australia), and it is positively correlated with latitude ($r_s = 0.32$; $P < 0.05$). Size at hatching is directly proportional to egg size in IP *Conus* species, and this is likely to be the case elsewhere in the world as well, but of course we have no direct evidence, except for the three temperate Australian species studied by Kohn (1993).

Whether or not the life history includes a planktonic larval stage is known for several extra-IP species even though the size of their eggs or hatchlings is not (Table 6.1). Although the sample sizes are small and include both tropical and temperate latitudes, the proportion of species with non-planktonic development is clearly higher outside (about 60 per cent) than within the IP region (less than 20 per cent).

7 DISCUSSION, SYNTHESIS, AND CONCLUSIONS

More than 250 species of *Conus* likely inhabit the vast tropical Indo-Pacific region. Our analyses of information from 62 of these indicate that several striking patterns and gradients occur among the quantifiable aspects of their reproductive biology.

In these analyses, we treated the reproductive and life history traits of *Conus* species as independent data points. But all species are not equally distinct phylogenetically, and since the seminal paper of Felsenstein (1985), the new approach of incorporating information on relatedness of taxa has strengthened inferences in this and all other types of comparative biological studies (see also Harvey and Pagel 1991). Unfortunately, although *Conus* has arguably the most interesting evolutionary history of any marine animal, no phylogenetic analysis has yet been attempted, and we are just now in the process of establishing a satisfactory character analysis (Röckel *et al.* 1993; Kohn, in prep.).

Despite this absence of inferred phylogenetic patterns in *Conus*, several aspects of this study strengthen the interpretations of the results. First, all comparisons are among species within one genus, and they 'are thus particularly informative because they automatically hold constant all the variables that are shared by congeners' (Harvey and Pagel 1991). Secondly, the results span 62 species, suggesting that similar correlations among character sets occur in independent lineages within the genus, and 'this means the traits have tended to evolve in a correlated fashion and explanations associated with phylogenetic history are unlikely to apply' (Harvey and Pagel 1991). Thirdly, as indicated below, striking correlations of reproductive and life history characters with the current, albeit phenetic, infrageneric classification are absent.

The simplest as well as the most critical variable in *Conus* reproductive biology is egg size, which ranges remarkably broadly (125–1000 μm diameter) among species within the genus. This is more than half the known range of egg size in all prosobranch gastropods (50–1500 μm), but very few of these exceed 1 mm (Perron and Carrier 1981; Fretter 1984). To the authors' knowledge, only the family Turritellidae is known to embrace a broader range of egg size among prosobranchs (100–1000 μm; Bieler and Houbrick 1990).

Although the evidence presently available is entirely descriptive-correlative, egg size appears to control several other reproductive and early life history traits. Maternal investment in offspring, duration of the prehatching development period, size and shape at hatching, developmental mode, larval growth rate, and duration of the precompetent planktonic period can now all be predicted with confidence from egg size in *Conus* (Perron 1981*a,b,c*; Perron and Kohn 1985; this monograph). Studies on other molluscan groups have yielded similar but much more limited correlations, e.g. bivalves (Sastry 1979); opisthobranchs (Hadfield and Switzer-Dunlap 1984); *Crepidula* (Hoagland 1986); temperate Western Atlantic prosobranchs (Lima and Lutz 1990). Among non-molluscan marine invertebrate taxa, egg size is also highly significantly correlated with developmental mode in the phylum Sipuncula (Rice 1989), but the largest body of data is for echinoderms. Comparisons of similar species from the same locations, and across the classes Echinoidea and Asteroidea, indicate similar patterns (Emlet *et al.* 1987); we discuss some specific aspects below. In addition, similar relationships between egg size and fecundity, hatching size, and larval duration characterize the fishes that share coral reef habitats with *Conus* (Thresher 1987).

The correlations listed above suggest the hypothesis that egg size is the factor controlling other life history parameters. This hypothesis would be supported by independent evidence that egg size is subject to natural selection free of phylogenetic constraints. Although as noted above the phylogeny of *Conus* species is unknown, their classification into phenetically based infrageneric groups permits limited evaluation of the relationship of egg size to taxonomic affinity. The results for five infrageneric groups (Table 4.1) indicated that egg size differs significantly in four of the ten possible pairwise comparisons. The only pattern in the data is that eggs of the *C. textile* group (11 species) are significantly larger than those in three other infrageneric groups.

In an adroit attempt to determine the degree of independence of egg size from phylogeny in echinoderms, Lessios (1990) compared egg size between congeneric geminate species isolated by the isthmus of Panama and between different genera on the same side of the isthmus. His results indicate that egg sizes of congeners have diverged by up to 36 per cent in linear dimension between Atlantic and Pacific populations separated for about 3 million years. The pattern in six of seven geminate pairs of larger eggs on the Atlantic side is consistent with adaptation to conditions of lower primary productivity there. While lineage-dependent differences were found within each ocean, egg size in echinoderms does appear to be subject to adaptational shift during a few million years of independent evolution. Many extant *Conus* species that have a fossil record are known to be 3–8 million years old (Kohn 1985). It is reasonable and perhaps likely that the egg size differences we describe here between species of *Conus* have also been independently affected by natural selection in response to environmental factors, and by constraints imposed by phylogeny at the level of the infrageneric group.

Concluding causation from correlations is unjustified, and the associations we have observed between egg size and other life history attributes could be due to relationships with other traits (McEdward 1988). The strongest support for the hypothesis that differences in egg size actually cause differences of the types reported here in the form and timing of early development comes from recent experimental studies of both invertebrates and vertebrates. Sinervo and McEdward (1988) reduced the size of developing echinoid eggs by excising blastomeres. Experimentally reducing *Strongylocentrotus droebachiensis* eggs to the size of the normally smaller eggs of its congener *S. purpuratus* resulted in smaller larvae with a simpler body form resembling that of *S. purpuratus*. The artificially reduced larvae also developed more slowly through the early feeding stages, but size at metamorphosis was independent of egg size. Hart (1992) confirmed these results and showed that reduced larvae feed at a lower rate, but found that their juvenile rudiment and size at metamorphosis were smaller than normal and depended on food available to the larva.

In the lizard *Sceloporus*, Sinervo and associates have experimentally both reduced and enlarged egg size, resulting in comparable effects on life history attributes. Smaller eggs have a shorter incubation time and hatch at a smaller size, although hatchlings grow faster than their normal siblings for the first few weeks. Experimentally reduced egg size also reduces survivorship as well as fitness-related attributes such as sprint speed in *Sceloporus* (Sinervo 1990; Sinervo and Huey 1990; Sinervo *et al.* 1992).

The results of the echinoid studies exactly parallel the observed differences among *Conus* species with different egg sizes, and the results of the reptile studies are essentially similar. Such experiments might be expected to be difficult or impossible with the determinate eggs of molluscs. However, the fact that the first two cleavages in *Conus* are equal (Fig. 3.9) suggests that they might indeed be feasible. In a study of dorsal–ventral axis determination in the similarly developing eggs of the opisthobranch, *Haminoea*, Boring (1989) successfully ablated 3B macromeres with a laser microbeam. These embryos subsequently hatched as normal, but smaller, juveniles.

Egg diameters of IP *Conus* range from 125 μm to 850 μm, or over a factor of about 6.8. This is equivalent to a more than 300-fold range in volume, from 0.001 mm^3 to 0.32 mm^3. *A priori*, energetic investment in the egg itself, and thus the amount of material and energy available for prehatching or intracapsular development of the egg, is likely to scale with mass or amount of organic material.

Strathmann and Vedder (1977) demonstrated that this is the case for a variety of marine invertebrates with eggs up to 200 μm in diameter, although organic matter scaled with egg volume to the 0.75 power rather than linearly. In echinoderms with larger eggs but also with planktotrophic larvae (those with egg diameter up to about 300 μm), the density of organic matter does not decrease with increasing egg volume, and it increases in the largest eggs of lecithotrophic species (Turner and Lawrence 1979; Emlet et al. 1987). Organic matter may thus be more concentrated in the smallest and largest eggs than in those intermediate in size.

The known range of egg diameter among *Conus* species world-wide is even greater—to 1 mm, or over a factor of 8 (equivalent to a 500-fold range in volume)—when the South African *C. tinianus* and the temperate Australian *C. papilliferus* are included. This represents more than half of the entire range in egg diameter reported for all prosobranch gastropods (60–1700 mm; Amio 1963; Fioroni 1966). It probably exceeds that of any other molluscan genus. Spight (1976) reported an order-of-magnitude range in *Thais*, but this muricid genus has subsequently been divided, and the maximum range of egg size in currently accepted genera (Kool 1987) is 590–920 mm in *Nucella*. The second most extensively studied gastropod genus appears to be *Crepidula* (Fioroni 1966; Hoagland 1986). The latter author reports egg diameters to range from 136 mm to 440 mm in 19 populations representing at least 16 species. The family Turritellidae is known to have a larger range of egg sizes (100–1000 mm) based on only three species in different genera (Bieler and Hadfield, 1990).

Although developmental modes in *Conus* cover both planktotrophy and lecithotrophy, the frequency distribution of *Conus* egg diameters is rather continuous and unimodal with positive skew (Fig. 4.1). This pattern contrasts with the bimodal distribution with a large gap between 250 and 350 μm previously thought to characterize the genus (Kohn 1961b), when fewer data were available. The form of Fig. 4.1 agrees with that predicted by the Perron and Carrier (1981) model of the relationship of reproductive efficiency and egg size for species with an encapsulated, benthic, pre-feeding stage, in which maternal energetic allocation to protective capsules increases with increasing size and decreasing number of eggs. Thus large eggs receive more parental care than do small ones. Perron and Carrier (1981) presented data from broad taxonomic groups (Muricidae, Prosobranchia, Opisthobranchia, and Bivalvia) in agreement with the model. The data presented here for the species in a single diverse genus provide an example of conformity with this model at a finer taxonomic level.

In a single spawn mass, the eggs of *Conus* are distributed among up to nearly 100 protective capsules. Depending on the species, all of these are affixed to a hard substratum, usually the underside of a rock (Type I); or some are affixed to a hard substratum and the remainder attached to previously laid capsules to form a cluster (Type II); or the first few capsules lack eggs and are embedded in sand as an anchor, to which the remaining egg-containing capsules are attached (Type III). Most Indo-Pacific species produce Type I egg masses, and Type III is the least common, presently being known only in *C. figulinus*.

Capsules in Type II egg masses are often thicker and tougher than those of Types I and III. This, combined with the tighter packing of Type II egg masses could impede water circulation through the egg mass and affect the respiration of the developing embryos. In contrast, egg capsules in Types I and III egg masses tend to be both thinner-walled and more widely spaced. Although nothing is yet known of water flow patterns near and around any *Conus* egg capsules, this seems unlikely to limit respiration of embryos. In fluid-filled capsules, ventilation depends more on circulation of the fluid within than on flow around the capsules (Strathmann and Chaffee 1984), and the flattened shape of *Conus* capsules reduces the distance across which oxygen must diffuse to the innermost embryos. In contrast to, for example, the gelatinous masses of adherent eggs of opisthobranchs as well as some polychaetes and fishes, oxygen consumption by embryos in flat, fluid-filled capsules can increase in proportion to capsule surface area and would not be expected to limit embryo size (Strathmann and Chaffee 1984). As expected from this hypothesis, egg size in *Conus* is unrelated to egg capsule size (Chapter 4).

The number of eggs deposited within a single *Conus* egg capsule varies much more widely among species than does egg diameter. In this study it ranged over more than 3 orders of magnitude, from 10 (*C. furvus*) to 50 000 (*C. vexillum*). For the genus as a whole, the minimum number of eggs per capsule known is 4–5, in the very small, recently described *C. diminutus* (adult shell length to 19 mm) from the Cape Verde Islands (Trovão and Rolán 1986), and *C. vexillum* has the maximum. Among prosobranch gastropods in general, many species deposit a single egg in each capsule (Fioroni 1966: Table 17). The maximum reported for the subclass appears to be the estimated 100 000 eggs per capsule in *Volutopsis norwegicus* (Gmelin) (Thorson 1940*b*). Almost all of these are nurse eggs, however, and they are devoured before hatching by the 1–4 embryos that complete development. No species of *Conus* is known to produce nurse eggs. Virtually every egg in each capsule is fertilized and undergoes development. This proved to be particularly favourable for our studies, because the presence and consumption of nurse eggs greatly increases the variance of some early life history attributes and thus precludes 'the potential for prediction of hatching size, developmental mode, and fecundity which are often made on the basis of taxonomy, egg size, or adult size' (Hadfield 1989).

In *Conus* as in other invertebrates, egg size and number are strongly and inversely related. The power function $Y = 1.6 \times 10^{10} X^{-3.0}$, where Y = number of eggs per capsule and X = egg diameter in μm, explains 72 per cent of the interspecific variance in egg number. This fit and the exponent of -3 (the RMA slope is -3.5) suggest a linear relation between egg number and egg volume or mass rather than egg diameter. The linear regression of egg number on egg diameter (Fig. 4.6) explains only 13 per cent of the variance. This reflects the result previously demonstrated for 10 Hawaiian species by Perron (1981*b*) that species with small eggs allocate relatively much less energy to protective capsules than do species with large eggs.

The total number of eggs in an egg mass, calculated as the product of the mean number of eggs per capsule and the number of capsules, only roughly estimates fecundity in *Conus*, for several reasons: it was feasible for us to count relatively few capsules in each egg mass, especially in those with very many eggs per capsule; between-capsule variation was high; also the discovery of an ovipositing female inevitably disturbs her enough to interrupt oviposition, and this causes the number of capsules to be underestimated.

Moreover, the number of times a female spawns in a year or in a lifetime is unknown in most species of *Conus*. In nature, a female *C. pennaceus* spawns once per year (Perron 1983). Repeated examinations of 52 individually marked females that spawned in April–June 1978 indicated that none spawned a second time that year. In 1979, 21 of these individuals were observed to spawn again. In the laboratory, two spawnings 4–5 months apart were observed in 3 of 20 *C. pennaceus* kept with unlimited food; Perron (1983) suggested that this was likely an artifact of laboratory conditions. Zehra and Perveen (1991) also observed spawning two or three times during the same breeding season in the laboratory in *C. coronatus*. Penchaszadeh (1984) reported that one *C. spurius* female laid two egg masses in an aquarium in Venezuela; he did not state the interval between them.

Throughout the genus in the IP region, according to our best estimates, the number of eggs per egg mass also varies over 3 orders of magnitude, from fewer than 10^3 to 1.5×10^6 (Appendix Table 1). Only *C. ventricosus*, the only representative of the genus in the Mediterranean, appears to produce a clutch of fewer than 100 eggs. The numbers of egg capsules produced by eastern Atlantic and South African species with very few eggs per capsule remains completely unknown; these species may well also produce small clutches (Appendix Table 2).

The relationships of adult size of *Conus* with the reproductive characteristics summarized above are complex. The body size of animals profoundly influences many aspects of their biology, including reproduction. Egg size as well as total egg production may vary directly with female size, as in the coho salmon (van den Berghe and Gross 1989), tropical angelfishes (Thresher and Brothers 1985), and some turtles (Congdon and Gibbons 1987). In other oviparous vertebrates, egg size and other components of reproductive effort per offspring do not vary significantly with the mother's size. Models that predict optimization of fitness of parents are based on the consequences of reciprocally varying egg size and number (Smith and Fretwell 1974), but number is usually more likely to be more sensitive to the factors influencing reproductive effort. 'In relatively stable environments, size of individual offspring should most often be under strong normalizing selection that reduces variation in egg size and offspring size' (Congdon and Gibbons 1987). In such environments, animals that produce large numbers of offspring that are given little parental care are thus expected to vary their reproductive output by varying the numbers rather than the sizes of eggs.

In marine invertebrates, egg size is typically assumed to be a species characteristic that does not vary systematically with female size. This conforms to the model just described, but explicit evidence is wanting. Sampling spawn masses from several mothers of the same species that varied in body size enabled us to test this assumption in 11 *Conus* species (Table 3.2; Figs. 3.2, 3.12). Among the eggs within single egg masses, the median coefficient of variation of egg diameter was 3 (range 0–10; $N = 97$ egg masses; Appendix Fig. 1). Among individuals of the same species, the median CV of egg diameter was 6 (range 1–23; $N = 35$ species; Appendix Fig. 1). These ranges of within-individual and within-species variation are quite similar to those reported for echinoids (summarized by Emlet *et al.* 1987).

Except for one species in which egg diameter varied significantly and inversely with adult size, rank correlation coefficients ranged from −0.33 to +0.46, and none was significant. We conclude that egg diameter is generally constant within species of *Conus*, regardless of adult size, and that it may thus be considered a species-level character.

Egg size and adult size in *Conus* are also unrelated on an interspecific basis. The only pattern in the scatterplot (Fig. 4.4) is that all species of large adult size (shell length > 60 mm) have egg diameters less than 420 μm. A similar absence of relationship between egg size and adult size occurs in Sipuncula (Rice 1989) and among 24 species of Central American echinoderms (Lessios 1990). Among teleost fishes, maximum egg size tends to increase with body size, but many large species have very small eggs (Duarte and Alcaraz 1989). In contrast to a common trend among marine invertebrates for the smallest species of a taxon to produce few, large, lecithotrophic eggs, to increase parental care, and to reduce planktonic larval duration (Valentine and Jablonski 1983; Strathmann 1990), there is no tendency for small species of IP *Conus* to have large eggs. In the eastern Atlantic region, the small species studied by Rolán (1985, 1986a) and Trovão and Rolán (1986) do produce very small egg capsules containing very few rather large eggs (Chapter 6).

Our data are not adequate to determine for any species whether larger females produce more egg capsules per clutch. Fecundity as estimated from clutch size does scale to adult size, however, because larger females make larger capsules—the range is a factor of about 6 (Fig. 4.5)—and larger capsules tend to contain more eggs. These relationships hold both among species (Figs. 4.5, 4.6A) and over a striking size range of reproductive females within species (Table 3.2; Figs. 3.2, 3.12). Studies of other marine prosobranchs have also shown that egg capsule size and number of eggs per capsule increase similarly with adult size among species of Muricidae (Spight *et al.* 1974; Rawlings 1990) and in the turrid *Oenopota levidensis* (Shimek 1986). In the vermetid *Petaloconchus montereyensis*, egg capsule size and number of eggs per capsule are larger in a northern (Washington) population of larger adults than in a southern (California) population having smaller adult size, while egg size and number of eggs per capsule are not. Nurse eggs occur in this species, however, and the one successful developing juvenile per capsule emerges at a larger body size in the northern population (Hadfield 1989).

With respect to growth pattern, *Conus* appears to be an 'intermediate strategist' (Sibly *et al.* 1985), allocating considerable energy to growth after attaining reproductive size rather than stopping growth after first breeding and devoting maximum available energy to reproduction. Among females of the same species, the shell length of the largest reproductive female was 1.2–3.0 times (average 1.8×) that of the smallest observed (Table 3.2; Figs. 3.2 and 3.12; Perron 1982: Fig. 5). Females clearly appear to continue to grow substantially after becoming sexually mature, although we have not been able to prove this by following the reproductive history of individual females. This would likely require several to many years, judging by the growth curves that Frank (1969) derived for two species. If the intermediate strategy is assumed based on the data in Figs. 3.2 and 3.12 and Appendix Table 1, it contrasts with the pattern in another family of prosobranchs, the Muricidae, in which adults grow little if at all (Spight *et al.* 1974).

According to the theory developed by Sibly *et al.* (1985), growth after breeding reduces fitness, because the energy devoted to subsequent growth could have been used for reproduction, unless certain conditions are obtained. Growth after breeding can increase the fitness of mature females if fecundity increases or mortality decreases per unit size. We lack information on size-specific mortality rates in *Conus*, and our results indicate that reproductive effort per unit size does not increase with increasing body size, either within species (four species reported by Perron 1982, and *C. textile* noted above), or across species (Fig. 4.12).

Another hypothesis advanced to explain the maintenance of an intermediate strategy is that a size constraint prevents females from investing more than a certain amount of energy into reproduction (Perrin *et al.* 1987). This hypothesis also leads to the prediction that the proportion of energetic effort devoted to reproduction should increase with increasing size, which as noted above is not the case in *Conus*. This model might be expected to be more applicable to species in which mothers brood or guard their offspring, for example many mesogastropods.

As expected from the observation that small eggs have short embryonic developmental periods (Fig. 4.7), they also hatch as small larvae. For the 42 *Conus* species analysed, the OLS regression of hatching size (Y, maximum dimension of shell in mm) on egg diameter (X, in μm) $Y = 0.002X - 0.144$ explains 93 per cent of the variance in the dependent variable (Fig. 4.9). This pattern conforms to that in neogastropods in general, for which Spight (1976) reported a similar regression but with a steeper slope, $Y = -0.014 + 0.019X$ ($r^2 = 0.84$).

Spight (1976) argued from optimality premises that the size of hatchling gastropods and their survival are likely to be directly related. He proposed three predictions from this hypothesis:

(1) hatching size should be larger in harsher environments;

(2) because related species have similar physiological tolerances and ecological requirements, they should have similar hatching sizes;

(3) high fecundity and large hatching size can both be attained if tissue density can be decreased.

Our data set permits a detailed examination of the second prediction. For the Neogastropoda as a whole, hatching size ranges from about 0.2 to 1.0 mm (median = 0.5 mm) for species with planktonic larvae, and from about 0.8 to 35 mm (median = 1.6 mm) for species with non-planktonic development (Spight 1976: Fig. 1C). The maximum shell dimension at hatching in 35 species of *Conus* with planktonic larvae varied from 0.24 to 0.84 mm, and non-planktonic larvae are usually about 1.25 mm long at hatching. Thus the predominantly planktonic larvae of *Conus* species vary in size over nearly the entire range of the suborder Neogastropoda. Egg size appears to be a more important correlate of hatching size than taxonomic affinity, and species of the genus *Conus* thus do not confirm the phylogenetic constraint of Spight's second prediction.

Egg diameter is a reliable predictor of the durations of embryonic development within the egg capsule (prehatching period) and of the minimum planktonic, planktotrophic period required prior to metamorphosis of the larva (precompetent period). In 37 egg masses of 29 species followed from oviposition to hatching, the prehatching period ranged from about 8 to 25 days (mean = 14 days).

We do not know either how long the process of oviposition of a single clutch takes, or how long the larvae take to emerge from their capsules at the end of the embryonic period. If the average time that Bandel (1976) observed in *C. jaspideus* holds generally, the oviposition rate would be about 3 capsules per hour, and most *Conus* egg masses could be deposited within 24 hours of continuous effort. The duration of the hatching process from a single egg mass could be an important aspect of early life history. Spawning is an infrequent event for a *Conus* female, and asynchronous hatching of the capsules of a single egg mass over an extended period could spread sibling larvae more broadly (Strathmann 1974).

Precompetent periods of 11 species have been determined under constant conditions in the laboratory (Perron and Kohn 1985). The mean precompetent duration is 21 days and the maximum, in the species with the smallest eggs, exceeds one month. The OLS regression of these durations on egg diameter is superimposed upon the data for species with known prehatching periods in Fig. 4.7, and it is used to estimate precompetent periods of the species in which these were not determined experimentally. With increasing egg size, the precompetent period tends to decrease faster than the prehatching period increases (Fig. 4.7), so that total development time to metamorphic competence (indicated by the total height of the histograms in Fig. 4.7) is shorter for species with larger eggs, up to about 470 μm. This is true for the entire observed data set of species with planktonic larvae (Fig. 4.7, with estimated data omitted; ANCOVA: $P < 0.001$), and for the comparison between species with planktonic larvae vs. those with eggs larger than about 470 μm, which have non-planktonic development (Mann–Whitney U test: $P < 0.02$).

This observed pattern conforms with the prediction of the Vance (1973) model of reproductive strategies that larger eggs develop to metamorphosis more rapidly than small eggs, which develop into feeding larvae that require more time to attain the size necessary for metamorphic competence.

In the only other extensive within-genus comparison that has come to our attention, Hoagland's (1986) study of 18 species of *Crepidula* demonstrated that all species with non-planktonic development, excluding those with nurse eggs, had larger eggs than those that hatched as planktonic veligers. As in *Conus*, the planktonic veliger larvae of *Crepidula* and other Calyptraeidae are planktotrophic; they use the velum as a ciliary food-collecting organ. Spight (1976) noted a similar relation but with slight overlap in the family Muricidae (*sensu lato*). Only two of 11 species (in nine genera) with non-planktonic development had smaller eggs than the largest of 16 species (in 11 genera) that hatch as planktonic veligers.

Among the seven species of *Crepidula* that hatch as planktotrophic veligers and for which both egg diameter and size at hatching are known (Hoagland 1986; Lima and Lutz 1990), hatching size is positively correlated with egg size ($r_s = 0.38$) but not significantly so ($r_{s.05} = 0.71$). By combining data from four species from two prosobranch families with planktonic development and four species from four families with non-planktonic development, Lima and Lutz (1990) reported a significant correlation of egg diameter with hatching size. However, this relationship is due to the difference between the groups representing the two developmental modes rather than significant correlations within each group.

The most thorough studies of these aspects of invertebrate reproduction and larval biology are for echinoderms, where a similar increase in time of development to metamorphosis with smaller egg size has been documented among co-occurring echinoid species (McEdward 1984), in eggs experimentally reduced in size (Sinervo and McEdward 1988), and statistically among echinoids ($r_s = -0.39$; $P < 0.05$) and among asteroids ($r_s = -0.36$; $P < 0.05$) (Emlet et al. 1987). These correlations are based on far more heterogeneous data sets than ours for *Conus*. The 34 echinoid and 38 asteroid species vary widely in degree of taxonomic relatedness within their classes, and in the range of latitudes and temperatures of their environments. Within a given range of temperature or latitude, however, total development from egg to metamorphosis takes longer in planktotrophic than in lecithotrophic asteroids (Emlet *et al.* 1987).

The planktonic larvae of all *Conus* species that have been reared in the laboratory are obligately planktotrophic, that is they must feed on planktonic microorganisms or particles and grow in order to attain competence to metamorphose. An egg diameter of at least 470 μm appears to be necessary to support non-planktonic development. In such species, veliconcha larvae capable of crawling typically hatch from the egg capsule. They retain the velar lobes and may swim, but for less than a day. It is likely but not yet demonstrated that development in these species is lecithotrophic, that is the larvae survive through metamorphosis on yolk supplied by the mother and they do not need to feed. Possibly they are lecithotrophic but may be able to ingest and assimilate phytoplankton, as in some nudibranchs. Facultative feeding by the latter larvae may extend the planktonic period or reduce metabolic loss of tissue during development (Kempf and Hadfield 1985; Kempf and Todd 1989). The latter authors note that their results demonstrate 'clearly the absence of a *simple* planktotrophic/lecithotrophic dichotomy' in opisthobranchs, and the same may well be true among species of *Conus*.

Kempf and Todd (1989) also review the integrated complex of characteristics required for successful lecithotrophic development. These are associated with:

(1) lengthening the prehatching embryonic period;
(2) shortening the larval developmental period, as demonstrated above for *Conus* (Fig. 4.10).

Some but not all of the specific characteristics listed by Kempf and Todd (1989) have been demonstrated to occur in *Conus* (Table 7.1).

Information on the duration of the planktonic competent period in *Conus* is completely lacking. On theoretical grounds, Jackson and Strathmann (1981) proposed that the durations of the precompetent and competent stages of inshore marine invertebrates that cannot control their horizontal transport should be directly correlated, and Pechenik (1980) presents evidence that this is approximately the case in three gastropods (*Ilyanassa obsoleta*, *Crepidula fornicata*, and *Bittium alternatum*). In the laboratory, he was able to extend the competent period 1–2 months, 3–4.5 times the duration of the precompetent stage. During the extended competent period, the larvae continued to grow at rates similar to their precompetent rates. In a fourth species, *Crepidula plana*, the maximum observed duration of the competent stage also exceeded the precompetent duration, but only by 10–25 per cent (Lima and Pechenik 1985).

The absolute size of eggs at the switch point from planktonic to non-planktonic larvae is larger in *Conus* than for many other invertebrates, where it is often 200–350 μm or less (Thorson 1950; Hermans 1979; Emlet *et al.* 1987). In *Crepidula*, species with eggs 136–260 μm in diameter have planktonic veligers (unless nurse eggs are present); those with eggs 300–440 μm in diameter hatch as crawling, benthic pediveligers (Hoagland 1986). As in *Conus*, the planktonic veliger larvae of *Crepidula* are planktotrophic. At least one species at the lower end of the lecithotrophic range (*C.* sp. cf. *C. convexa* from Florida; egg diameter 230 μm) hatches as a veliconcha that feeds with the velum, while the benthic hatchlings of those that develop from larger eggs feed only with the radula (Hoagland 1986).

In contrast to size at hatching, size at settlement is not correlated with egg size in the 20 species of *Conus* with available data (Fig. 4.10). Although the maximum linear dimension at hatching ranges over nearly an order of magnitude among species, at settlement it varies only by a factor of 2. The growth curves of planktotrophic larvae demonstrated by Perron (1981*a*) in six Hawaiian species indicate that those hatching at small size grow rather slowly during their long planktonic phase, while

Table 7.1 Kempf and Todd's (1989) tabulation of 'Changes in embryonic and larval characteristics during a transition from obligate planktotrophy to obligate lecithotrophy', applied to Indo-Pacific *Conus*

Characteristic	Demonstrated in *Conus*?
A. Longer embryonic developmental period leading to a more developed larva	yes (Fig. 4.7)
1. Per embryo (a) increased yolk storage or (b) uptake of DOM	(a) yes
2. Increased durability of egg mass	yes (Perron, 1981*b*)
3. Changes in hatching enzyme composition and time of release	? (likely)
B. Shorter larval developmental period	yes (Fig. 4.7)
1. Incorporation of larger yolk-derived energy reserves (or increased DOM)	? (very likely)
2. Greater reliance of larval metabolism on yolk-derived energy (or DOM uptake)	
(a) Reorientation of Golgi-endoplasmic reticulum-lysosomal systems toward yolk reserve storage and use (or DOM uptake)	?
(b) Loss of ability to digest ingested food	? (unlikely; velum persists)
(c) Loss of unnecessary digestive and feeding organs	? (unlikely)
3. Loss of inducer-mediated metamorphosis	? (unlikely; absent in planktotrophic larvae)
4. Loss of larval stage; development direct	no (possible in some non Indo-Pacific species)
5. Further reduction of 'larval' characteristics in embryo	?

those hatching as large veligers grow rapidly during their shorter planktonic phase. These trajectories tend to converge on a common size at competence, and they are consistent with the sparser data on hatching and settlement sizes of the other species presented here. These results suggest that there is a critical minimum size for settlement and metamorphosis in *Conus*, usually in the range of 1.0–1.5 mm, regardless of egg size and hatching size. This interpretation is consistent with Christiansen and Fenchel's (1979) observation that size at metamorphosis is generally rather constant within genera of prosobranch gastropods.

Taylor's (1975) observations of the somewhat larger settlement size of larvae of seven additional species collected from the plankton in Hawaii (mean 1.7 mm, range 1.2–2.2 mm) suggests that growth likely continues during the competent period. While small size may be adaptive to the planktonic larva which must stay afloat and depends on collecting small food particles with cilia, a minimum size for settlement and metamorphosis is likely to be associated with the need of the juvenile to adopt a predaceous habit immediately. The smallest post-metamorphic juveniles of *Conus* examined are known to prey on polychaetes (Taylor 1975; Nybakken and Perron 1988).

In an analysis of developmental and life history characteristics of opisthobranchs, the only other similar data set for gastropods known to the authors, Hadfield and Miller (1987) also found size at settlement to be uncorrelated with size at hatching among the 25 species with available data. In this subclass, egg diameter (40–380 μm) and hatching size (100–280 μm) are generally smaller than in prosobranchs, and size at settlement (150–500 μm) is as variable as hatching size. In polychaetes, echinoderms, and some crustaceans, as in *Conus*, size at settlement is more uniform than either egg size or adult size (Strathmann 1977; Hermans 1979; Emlet *et al.* 1987). Among echinoderms, size at metamorphosis is not related to egg size in echinoids, but asteroids show a very different pattern. Egg size in this class is strongly bimodal, the modes corresponding to planktotrophic and lecithotrophic development. In the former group, with small eggs, size at metamorphosis is inversely correlated with egg size ($r_s = -0.65$; $P < 0.05$; $N = 23$), while in lecithotrophic species ($N = 14$) and in the class as a whole ($N = 31$), the correlations are positive ($r_s = 0.80$ and $r_s = 0.45$, respectively) and significant (Emlet et al. 1987).

Whether maternal investment in reproduction differs among *Conus* species with different egg sizes and developmental modes remains unknown. Hughes and Roberts (1980) found no such differences between *Littorina* species with planktonic and non-planktonic development, and they cited concurring evidence from a few other gastropod groups. Although the energetics of reproduction remain very poorly known in *Conus*, two patterns are fairly clear. First, in species characterized by producing a small number of large eggs that have a long intracapsular developmental period, females devote more energy to protective capsules than do females that lay large numbers of small eggs, which have shorter prehatching periods. Secondly, the volume of eggs in an egg mass does not scale linearly with body volume of mothers. The OLS regression of the logarithms of total egg volume on the logarithms of mothers' shell length among species (Fig. 4.12) is highly significant but with a coefficient of 2.4 rather than 3 (the RMA regression slope is 2.8), indicating that larger females invest relatively less energy in eggs than do smaller ones. Few comparable data are available for other prosobranch groups, but Spight and Emlen (1976) found a highly significant regression of clutch size (expressed as log egg number) on female shell length, with a regression coefficient of 5.7 ($r^2 = 0.93$). The independence of egg size from size at larval settlement and from adult size in *Conus* indicates that different selective factors influence body size at different life history stages.

Alternative hypotheses for the evolution of egg or offspring size are usually cast in the form of optimization models:

1. Size-specific vulnerability. If the egg and embryonic stages constitute a 'safe harbour', i.e., embryos suffer significantly lower mortality rates than do free-living juveniles of similar body size, selection will favour evolution of larger egg size (Shine 1978, 1989).

2. Coevolution of egg size and parental care. An increase in either variable initiated by some other factor favours an increase in the other (Nussbaum and Schultz 1989).

In (1), the factor selecting for large egg size, for example desiccation or predation, may simultaneously but independently select for larger eggs and increased parental care, but the latter two variables are themselves not causally related or genetically correlated and cannot coevolve (Shine 1989). If this were the case in the *Conus* species for which we have presented data on reproductive habits (Chapter 3), we would expect some correlation of egg size with habitat attributes. However, the eight species of IP *Conus* with eggs > 470 μm, and thus known or estimated to have non-planktonic development, occupy the same coral reef platform and subtidal sandy bay habitats as congeners with small eggs and long planktonic larval stages. All species known to occur on intertidal benches (e.g. Kohn 1987) where desiccation might be a relevant stress, produce small eggs and larvae with long planktonic stages. It thus seems that the observed positive correlation between egg size and parental investment in protective egg capsules in *Conus* is somewhat more likely to be coevolved than to result from independent but parallel selective factors.

Several models beginning with those of Vance (1973) have sought to estimate the number of successfully metamorphosing larvae using as input variables egg size and number, energy invested in eggs and protective capsules or other parental care, and mortality rate in the plankton. Perron and Carrier (1981) modified the model to accommodate the inverse relationship of energy devoted to egg capsules in *Conus* with increasing egg size and decreasing egg number. This model predicts the monotonic egg size to frequency distribution described above for *Conus* as well as in prosobranchs generally with encapsulated eggs. More recently Grant (1983) included a hypothesized reduction of mortality with increased energy devoted to brood protection as a parameter in the model. Although mortality rates of eggs, embryos, larvae, and juveniles in nature remain notoriously difficult to measure, these early life history stages are doubtless subject to various selective pressures that affect survivorship during developmental time. It is highly unlikely, however, that selection acts directly on dispersal of larvae. Rather, factors that reduce parental investment per offspring, or enhance growth and survivorship of larvae in the plankton, secondarily or indirectly increase passive dispersal as well. 'Large scale dispersal appears to be an accidental byproduct of life history adaptation, though this byproduct may have profound evolutionary consequences' (Emlet *et al*. 1987).

Long ago, Thorson (1950) demonstrated that a large majority (he estimated 80–85 per cent) of tropical shallow-water marine invertebrates broadcast planktonic larvae. The estimate for tropical IP *Conus* based on our results fit his estimate exactly: counting *C. pennaceus* twice because planktonic and non-planktonic development apparently occur in different regions, 82 per cent of *Conus* species (51/62) have a planktonic larval stage. The proportion appears to be much lower in other regions. In the tropical western Atlantic, the majority of species with known developmental mode are non-planktonic (6/10; Table 6.1). Combining data from all known extra-IP regions, both tropical and temperate indicates an even higher non-planktonic proportion, 14/21 or two-thirds (Table 6.1). Overall, the data suggest that latitudinal patterns in egg size and developmental mode may exist within *Conus*, and that they conform with the trends observed in broader taxonomic groups (Thorson 1950; Spight 1977; but see Jablonski and Lutz 1983) of increasing prevalence of non-planktonic development at higher latitudes.

The fates of planktonic larvae of marine invertebrates are also extremely refractory to observation. These larvae are dispersed over widely varying spatial and temporal scales. 'In some instances dispersal is only a few centimeters or meters; in others larvae may be transported thousands of kilometers across wide expanses of ocean. The time interval over which dispersal occurs may be a few minutes, a tidal cycle, many months, or even a year' (Scheltema 1986a). Developmental modes vary so widely across species of *Conus* that these broad scales are likely spanned within this one genus. Because the geographic ranges of IP *Conus* also vary widely—from single archipelagoes or small ocean basins or stretches of continental coastline to virtually the entire IP realm of some 10^8 km^2 of ocean area—examining the correlations of geographic distribution patterns with developmental modes enables a test of the dispersal theory of biogeography. This hypothesis predicts a positive correlation between dispersal ability and geographic range. The converse observation, species with excellent dispersal potential but narrowly restricted ranges, refutes the dispersal hypothesis. In such cases, the alternative hypotheses of vicariance or ecological determinism more likely explain the observed patterns.

Unfortunately, knowledge of *Conus* phylogeny has not yet progressed far enough to provide objective information on interspecific relationships that is required to test the vicariance hypothesis. However, a preliminary overview of the IP species suggests that several pairs of very closely related species (e.g. *C. ebraeus* and *C. chaldaeus*; *C. miliaris* and *C. coronatus*; *C. bandanus* and *C. marmoreus*) have very similar distribution patterns. Disjunct or allopatric distributions of most closely related species may not be a common feature of *Conus* biogeography. Clearly a major challenge for future research in this field is to relate objectively determined phylogenetic hypotheses to geographic distribution patterns. Given the present lack of such information, the predominant evidence favours the hypothesis that dispersal ability is an important determinant of species distributions. In particular, this evidence derives from analyses of three types of distribution patterns and the developmental modes of the species fitting them, as far as these are known.

First, both non-planktonic development and short

precompetent durations among species with planktonic larvae occur frequently in species whose geographic distributions are predominantly linear or one-dimensional, i.e., distributed along continuous continental coastlines. Almost all species inhabiting the archipelagoes of the predominantly two-dimensional tropical Pacific and Indian Ocean basins have planktonic larvae, and these take longer to develop metamorphic competence than those of the first group (Table 5.1).

Secondly, species that are confined to the continental islands and coastlines of the Eurasian lithospheric plate tend to have non-planktonic development, while almost all of those whose ranges extend to the islands of neighbouring oceanic plates have planktonic larvae. For example, the shallow-water *Conus* fauna of the Philippine Islands comprises about 100 species. Of these, 63 also occur on the Philippine plate to the east, and 60 of these extend their ranges to the Pacific plate oceanic islands farther eastward.

In our final and simplest test of the dispersal hypothesis, we disregarded specific regional aspects and examined only the patterns of absolute range areas occupied by IP *Conus* species with varying developmental modes. We determined these areas as minimum convex polygons circumscribing all reported locality records on maps. The OLS regression of geographic range area of IP *Conus* species on length of the precompetent planktonic period is highly significant; the latter accounts for 40 per cent of the variance in the former.

Species of *Conus* with greater potential for larval dispersal thus have significantly broader geographic ranges, and we conclude that the evidence from larval life history studies supports the hypothesis that larval dispersal is an important determinant of geographic distribution patterns.

It is certainly not the only determinant, however, and exceptions to this pattern occur. Examples are the three outliers in the lower right of Fig. 5.4: *C. abbreviatus* (see also Fig. 5.2E), primarily endemic to the Hawaiian Islands but also recorded from Johnston Island and Enewetak in the Marshall Islands; *C. biliosus*, a mainly Indian Ocean species occurring along continental shores and evidently with a disjunct distribution (Fig. 5.2F), and *C. suratensis*, restricted to the Indo-Malayan region. One species with a very long planktonic stage, *C. abbreviatus*, is essentially endemic to the Hawaiian Islands (Fig. 5.2E; a few isolated individuals have been found in the Marshall Islands). This is reminiscent of the endemic Hawaiian wrasse *Thalassoma ballieui*, which has the longest known realized larval duration of any tropical fish (Victor 1986). At the opposite extreme, *C. pennaceus* is widely distributed yet has virtually non-planktonic development in Hawaii. Its Type II egg masses are sometimes dislodged and have been observed floating at the sea surface, raising the likelihood of alternative modes of transport and colonization of new environments.

In general, for a given egg size and hypothesized dispersal ability, *Conus* species with Types II and III egg masses tend to have broader geographic distributions than those with type I egg masses (Fig. 5.4). The former two types are much more likely to be torn loose from the substratum without fatally injuring the contained embryos, suggesting that floating of egg capsules may provide an additional mode of dispersal in some *Conus* species. Rafting of adult *Conus* is extremely unlikely. The adults are large and heavy, and the long, narrow foot is not conducive to adhering to large floating objects for long periods. This contrasts markedly with tropical corals whose planula larvae may settle and metamorphose on floating pumice (Jokiel 1984) and small, easily transported molluscs such as the bivalve *Lasaea*, in which species that brood young to the benthic crawling stage are more widely distributed geographically than those that release planktonic larvae (Ó Foighil 1989).

Although the authors know of no data sets for other taxa of marine animals closely resembling our comparative data base for *Conus*, several complementary approaches for examining the dispersal hypothesis have produced relevant results. During the past decade in particular, studies of planktonic larval dispersal and geographic distribution have focused actively and productively on coral reef fishes. These fishes occupy the same habitats as *Conus* and utilize some of the same resources, so their comparative analysis is likely to be especially fruitful.

Because the realized or actual length of the pelagic larval stage of many coral reef fishes can be directly determined from daily growth lines and 'settlement marks' in otoliths, it is possible to test the dispersal hypothesis by examining the relationship between total pelagic duration and geographic distribution patterns. This approach is not technically possible in gastropods, where as yet we have no means of assessing the duration of the competent portion of the pelagic period.

In a large sample of 115 species representing 22 families of IP coral reef fishes, Brothers and Thresher (1985) showed that species with longer pelagic durations tend to have more extensive geographic ranges (the trend in their Fig. 1 is significant at the 0.005 level by the Olmstead–Tukey corner test). Their most striking result was that species with pelagic durations longer than 45 days ($N = 23$) were distributed much more broadly than those with shorter planktonic periods ($N = 92$) (χ^2 test: $P < 0.001$). Although correlations within these arbitrary groups were not significant, dispersal is clearly an important determinant of distribution in IP reef fishes. Moreover, species occurring in Hawaii and those extending their ranges across the East Pacific Barrier to eastern Pacific reefs had significantly longer planktonic durations (Hawaii: mean = 49 days, SD = 10; eastern Pacific: mean = 62 days, SD = 11) than those with ranges restricted to the western and central Pacific (mean = 29 days, SD = 10) (Kolmogorov–Smirnov tests, $P < 0.001$). Victor (1986) observed generally similar patterns within the family Labridae (wrasses).

In a more intensive study of 31 IP species in the family Pomacanthidae (angelfishes), areal extent of geographic range was not correlated with planktonic duration (Thresher and Brothers 1985). However, egg diameter was negatively correlated with actual pelagic duration in the 10 IP species ($r = -0.61$; $P < 0.1$) and 13 IP + Atlantic species ($r = -0.70$; $P < 0.01$) for which these authors had data. The situation in Pomacanthidae is complicated by two factors absent in *Conus*, a significant positive correlation between adult size and egg size in both the IP and IP + Atlantic samples, and a highly significant negative correlation between pelagic duration and adult size among all 31 species.

As in Pomacanthidae, there is no relation between planktonic duration and extent of geographic range among species of Pomacentridae (damselfishes), which have shorter and less variable precompetent periods and seem unable to delay metamorphosis after competency (Wellington and Victor 1989). Geographic ranges of pomacentrids tend to be smaller and endemism more frequent than in wrasses (Wellington and Victor 1989).

An independent but equally relevant approach to the dispersal question is to compare genetic similarity among close and distant populations. Rosenblatt and Waples (1986) compared genetic distances, based on protein electrophoresis of 38 loci, between Hawaiian and eastern Pacific populations of 12 trans-Pacific shore fish species, and between Atlantic and eastern Pacific populations of the two circumtropical members of their data set. Their results support a much closer genetic relationship of allopatric populations than would be predicted from the vicariance hypothesis. Nei's genetic distance D ranged from 0.01 to 0.06 (mean = 0.03) between the Hawaiian and eastern Pacific populations, while for the trans-isthmian populations of the circumtropical species, D values were 0.14 and 0.19. The former values are typical of those found between local populations of fishes in general and of many other taxa, while the latter more closely resemble results from congeneric species comparisons. Rosenblatt and Waples (1986) argue that had the IP and eastern Pacific populations been separated by the closure of the eastern Tethys barrier (12–14 mybp), their genetic distances should be greater than those of populations separated by the Panamic isthmus (3.1–3.6 mybp). The results strongly suggest subsequent gene flow between geographically widely separated populations of IP shore fishes.

Reduced dispersal and smaller geographic ranges are not the only consequences to marine animals with a short or no planktonic larval period. Strathmann (1990) listed numerous potential effects, including more regular recruitment and enhanced adaptation to local conditions, more inbreeding and reduced genetic variation within, but enhanced genetic variation between populations, more interaction among kin, and higher rates of speciation and extinction.

All of these effects are evolutionarily significant, but none has yet been investigated in *Conus*. All are also likely related to egg size, knowledge of which has already proved to be an important key to understanding reproductive energetics, larval biology, and biogeography in this large genus with striking interspecific variation in these attributes. Given the egg diameter of a species of *Conus*, one can predict with confidence such life history parameters as the time from oviposition to hatching, the size and stage at hatching, whether or not the larva will be planktonic and planktotrophic, the minimum planktonic period, and the extent of parental investment in offspring. In *Conus* these life history traits thus scale with size of the offspring, rather than size of the parents.

Although we have determined the egg sizes of some 61 IP *Conus* species, and 13 from other parts of the world, we still lack this basic datum for the majority of species in the genus. However, we have also demonstrated that if egg diameter is not known, it may be reliably estimated from the diameter of the protoconch or first whorl of a well-preserved adult shell. At present, we have information only on the size and general shape of larval shells, and for rather few species. In the future, detailed comparative analyses of larval shell morphology are likely to lead to additional insights into *Conus* life history and the control of its attributes.

The comparative studies we have presented here have allowed us to couple the morphological relationships of egg and larval characteristics with the dynamics of larval dispersal and thus to predict the attributes of geographic distributions of congeneric species that vary widely in developmental mode. We conclude that among gastropods of the genus *Conus*, egg size is the single most important attribute for understanding

(1) reproductive energetics;
(2) the temporal patterns of embryonic development and larval biology;
(3) dispersal potential, which is tightly linked functionally to (1) and (2) but is an evolutionary 'byproduct'.

We also conclude that egg size has important implications for life history evolution, population ecology, and biogeography. Knowledge of these attributes of such a large, widespread, and ecologically important genus contributes importantly to our understanding of the remarkably complex biotas of the tropical seas. This knowledge may also help us understand spatial and temporal patterns of speciation and extinction, the primary evolutionary events that determine the nature of the history of life.

APPENDIX

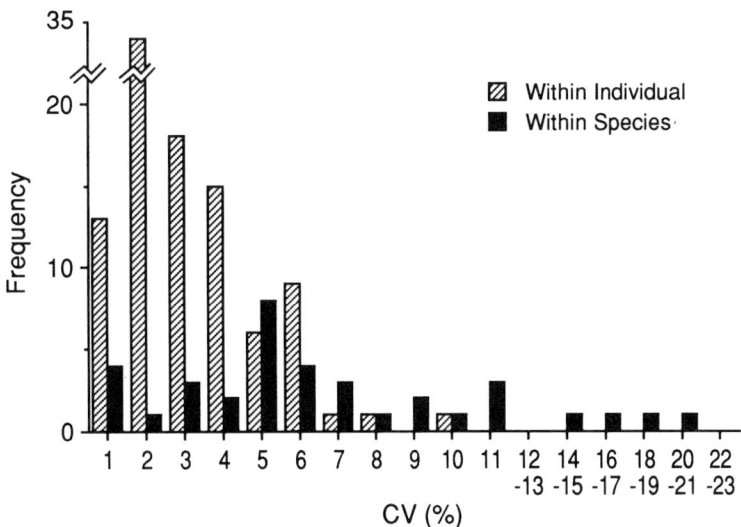

Fig. 1 Frequency distributions of coefficients of variation of diameters of *Conus* eggs. Hatched histograms, data for sibling eggs, i.e. from a single egg mass. Coefficients of variation (*V*) are uncorrected, as most *n* ranged between 10 and 16. *N* = 95. Solid histograms, data for mean egg diameters of conspecific mothers. Coefficients of variation ($V^* = V(1 + 1/4n)$ corrected for bias, as *n* ranged between 2 and 11. *N* = 36. Complete data are given in the Appendix Tables.

Appendix Table 1 Summary of data on reproduction and development of Indo-Pacific *Conus*

New record no.	Specimen no.	Species	Locality	Egg diam (μm) $\bar{x} \pm SD$	Eggs per capsule $\bar{x} \pm SD$	Range	Av. no. eggs per mass	Capsule size (mm) ht × max breadth	Prehatch period (days)	Hatchling size (μm)	Minimum pelagic period (days)	Settling size (SL, mm)	Adult size (mm)	References
		abbreviatus Reeve	Hawaii	—	—	—	—	10 × 8	—	—	—	—	—	Ostergaard (1950)
	1382–1383	*abbreviatus* Reeve	Hawaii	[170]	1300	—	44 000	9 × 7–7.5	14	280–300	[26]	—	29 × 21	Kohn (1961a)
		abbreviatus Reeve	Hawaii	170	—	—	—	10 × 9	14–15	270	32	1.12	33	Perron (1981a,b)
1	C158483	*achatinus* Gmelin	Australia	837	82 ± 38	32–122	—	15–17 × 10–12	—	—	[N]	—	52 × 24; 48 × 23	present study
2	C158484	*achatinus* Gmelin	Australia	854	34 ± 5	33–40	714	8–10 × 6–8	—	≥1490	[N]	≥1.49	36 × 17	present study
3	C158484	*achatinus* Gmelin	Australia	781	29 ± 2	27–30	781	8–11 × 5–8	—	—	[N]	—	—	present study
	3948	*ammiralis* Linnaeus	Marshall Is.	—	—	—	—	—	—	—	—	—	60 × 32	present study
1	FEP277	*ammiralis* Linnaeus	Marshall Is.	331 ± 5	271	—	—	10.5 × 9	13	660	[12]	—	51 × 27	present study
2	3898	*ammiralis* Linnaeus	Marshall Is.	348 ± 14	310	292–329	—	9.5–10 × 7–8.5	—	—	[11]	—	50 × 26	present study
3		*ammiralis* Linnaeus	Fiji	304 ± 17	521	510–533	—	12–15 × 8–10	—	—	[15]	—	—	present study
		araneosus [Lightfoot]	India	492	164	63–214	7400	9–17 × 5–11	—	1100	0–3	—	47 × 27	Natarajan (1957)
	JEN14	*araneosus* [Lightfoot]	Philippines	517 ± 9	43	30–55	—	12 × 10	—	—	[N]	—	40 × 23	present study
	5202	*arenatus* Hwass	Cosmoledo I.	—	1000	—	—	10–14 × 6–7.5	—	—	—	—	47 × 27	Kohn (1961b)
	5235A	*arenatus* Hwass	Assumption I.	—	—	—	—	13–13.5 × 10–11	—	>220	—	—	—	Kohn (1961b)
	5235B	*arenatus* Hwass	Assumption I.	—	4300 ± 954	3400–5400	65 000	10–11 × 9–10	—	—	—	—	—	Kohn (1961b)
	6147	*aristophanes* Reeve	Fiji	186 ± 9	1265	—	—	6.5 × 6.5	—	—	[25]	—	25 × 16	present study
	6319	*aulicus* Linnaeus	Indonesia	326 ± 7	1900	1590–2220	150 000	19–27 × 15–19	—	>440	[12]	—	96 × 40	present study
		balteatus Sowerby	Indonesia	311 ± 7	(434)	421–448	>10 000	11 × 10	—	—	[14]	—	37 × 23	present study
		bandanus Hwass	Hawaii	344	—	—	—	20 × 14	18.5–19.5	755	10	1.48	60	Perron (1981a,b)
	FEP280	*bandanus* Hwass	Palau	[299]	434	425–444	—	11 × 9	—	630	[15]	—	59 × 34	present study
		biliosus (Röding)	Pakistan	174	1300	1100–1300	27 320	7–10.5 × 6–8	11–12	251	[26]	—	—	Barkati & Ahmed (1985)
		biliosus (Röding)	Pakistan	172 ± 11	1192 ± 165	—	28 600	9–12 × 7–10	11	250	[26]	—	25–33	Zehra & Perveen (1991)
	5292	*canonicus* Hwass	Seychelles	—	1510	—	72 000	17–20 × 13–15	—	—	[21]	—	59 × 32	Kohn (1961b)
	5400	*canonicus* Hwass	Seychelles	233	1000	—	>15 000	11–11.5 × 8.5–9.5	14–15	365–400	[18]	—	42 × 20	Kohn (1961b)
1	6168	*canonicus* Hwass	Indonesia	260 ± 20	—	—	—	8 × 7	—	—	[18]	—	34 × 16	present study
2	6169	*canonicus* Hwass	Indonesia	258 ± 2	482	456–508	17 000	8–10 × 7	—	≥336	[18]	—	34 × 15	present study
3	6397	*canonicus* Hwass	Indonesia	270 ± 8	1400	1362–1438	76 000	13–14 × 10–12	—	417	[17]	—	48 × 24	present study
4	6912	*canonicus* Hwass	Indonesia	278 ± 12	654	637–688	24 000	11–12 × 8.5–10	—	—	[17]	—	42 × 20	present study
1	6032	*capitaneus* Linnaeus	Thailand	148 ± 6	3164	2818–3750	111 000	13.5–15 × 9.5–10.5	9(?)	209	[28]	—	39 × 23	present study
2	6033	*capitaneus* Linnaeus	Thailand	137 ± 5	3396	3357–3464	78 000	12–13 × 8.5–9.5	8(?)	—	[29]	—	37 × 21	present study

	Specimen	Species	Location												Reference
3	FEP30	capitaneus Linnaeus	Palau	161 ± 3	4675	4550–4800	—	15–16 × 12	—	—	[27]	—	—	44 × 25	present study
		catus Hwass	Hawaii	200	—	500–1000	15 000	12 × 9	15–16	375	[23]	—	—	—	Ostergaard (1950)
	2180	catus Hwass	Hawaii	220	1650	—	—	12 × 10	—	—	[22]	—	—	—	Kohn (1961a)
	2299	catus Hwass	Hawaii	235	1683	—	—	11 × 8.5–9.5	—	—	[20]	—	—	40 × 26	Kohn (1961a)
1	2298	catus Hwass	Hawaii	234	1604	—	—	11–12 × 9.5–10	—	—	[20]	—	—	—	present study
2	6069–6070	catus Hwass	Thailand	241 ± 6	—	115–910	18 000	8 × 6	—	—	[20]	—	—	31 × 17 (f); 30 × 18 (f)	present study
	JEN56	catus Hwass	Philippines	212	1532	1434–1633	—	13–16 × 9–11.5	—	—	[22]	—	—	40 × 23	present study
4	FEP53	catus Hwass	Palau	231 ± 4	783	750–816	—	10 × 7	—	—	[21]	—	—	36 × 17	present study
5	11455	catus Hwass	Niue	229 ± 10	1633	1400–1900	41 000	13–14 × 10.5–11	—	—	[21]	—	—	36 × 24	present study
	11456	chaldaeus (Röding)	Niue	175 ± 17	1252	1183–1320	—	8 × 7	—	—	[25]	—	—	30 × 19	present study
	JEN44	cinereus Hwass	Philippines	480 ± 10	173 ± 20	161–203	9700	11–12 × 7.5–9.5	—	—	[N]	—	—	51 × 25	present study
1	JEN18	cinereus? Hwass	Philippines	—	—	79–110	8900	9–10 × 7.5–9.5	—	≥756	[N]	—	—	42.5 × 19	present study
2	JEN57	cinereus? Hwass	Philippines	496 ± 20	100 ± 14			10–12 × 9–12	—	—	[21]	—	—	41 × 18	present study
3	FEP135	coffeae Gmelin	Palau	222 ± 6	475	450–500	—	7–8 × 5	—	—	[3]	—	—	29 × 15	present study
1	3946	consors Sowerby	Marshall Is.	434 ± 5	(157)	125–189	—	10 × 7.5	—	—	[3]	—	—	48 × 23	present study
2	3495	consors Sowerby	Marshall Is.	440	203	186–219	3650	1–12 × 10	—	—	[3]	—	—	40 × 20	present study
3	3348	consors Sowerby	Ponape	—	—	—	—	10.5–11 × 9–10	—	—	[4]	—	—	45 × 21	present study
4	JEN3	consors Sowerby	Philippines	419 ± 7	249 ± 60	179–319	8700	15 × 10–13	—	≥600	[4]	—	—	57 × 30	present study
5	JEN32	consors Sowerby	Philippines	409 ± 11	249 ± 15	231–258	—	12.5–14 × 10.5–13	—	—	[5]	—	—	54 × 27	present study
6	FEP35	consors Sowerby	Palau	—	—	—	—	—	—	—	—	—	—	42 × 21	present study
7	FEP36	consors Sowerby	Palau	389 ± 8	(213)	136–289	—	10–15 × 9–12	—	823 ± 41	—	1.46 ± 0.33	42 × 21	present study
8	FEP62	consors Sowerby	Palau	—	—	—	—	—	—	—	7	—	—	50 × 23	present study
	4092	coronatus Gmelin	Maldive Is.	165	700	760–850	52 500	4.5–5 × 4.2–5	—	240–250	[26]	—	—	14.5 × 8.5	Kohn (1961b)
	5218–5221	coronatus Gmelin	Cosmoledo	165	(805)	—	—	6 × 4	—	—	[26]	—	—	18–21 × 11–12	Kohn (1961b)
	4996	coronatus Gmelin	Seychelles	180	—	—	—	6 × 6	—	—	[25]	—	—	21 × 13	Kohn (1961b)
	5254	coronatus Gmelin	Seychelles	170	2633	2400–3000	>18 400	8.5–11 × 10–10.5	—	—	[26]	—	—	33 × 20	Kohn (1961b)
	5618	coronatus Gmelin	Seychelles	200 × 150–160	1800	2641–3309	25 200	8.5–10 × 7–10	—	—	[26]	—	—	29 × 18	Kohn (1961b)
	BRAR68	coronatus Gmelin	Pakistan	160	1535	1100–1940	36 840	7.5–13.5 × 6.5–11.5	10	215	[27]	—	—	—	Barkati & Ahmed (1985)
	C111634	coronatus Gmelin	Pakistan	151–169	1623 ± 300	—	36 800	9.9–12.5 × 9–12	8.5–9	216 × 160	[27]	—	—	—	Zehra & Perveen (1991)
	7384	coronatus Gmelin	New South Wales	165	1300	1875–2094	109 000	10–11 × 7–8	—	—	[26]	—	—	38 × 24	Huish (1978)
1	7385	coronatus Gmelin	India	166 ± 10	2011	2641–3309	89 000	9.5–10.5 × 5 × 9	—	—	[26]	—	—	33 × 20	present study
2	BRAR68	coronatus Gmelin	India	178 ± 9	2962	560–756	—	7–7.5 × 6.5–7	—	—	[25]	—	—	23 × 16	present study
3		coronatus Gmelin	Cook Is.	186 ± 13	638 ± 83	50–75	1250	8.5–10.5 × 6.5–7.5	—	—	[24]	—	—	39 × 18	Thorson (1940)
		dictator Melvill	Persian Gulf	575	62				—	≥1110	[N]	—	—		
1	FEP397	distans Hwass	Palau	148 ± 3	(5114)	5078–5150	36 800	15 × 12	1	—	[28]	—	—	84 × 46	present study
		ebraeus Linnaeus	New Caledonia	—	(2250)	1500–3000	54 000	7 × 6	>11	228–247	[26]	—	—	—	Risbec (1932)
		ebraeus Linnaeus	Hawaii	[167]	—	—	—	10 × 10	14	280	[26]	—	—	42	Ostergaard (1950)
		ebraeus Linnaeus	Hawaii	170	—	—	—	—	—	280	[25]	—	—	29 × 19	Perron (1981)
		ebraeus Linnaeus	Cosmoledo	180	1983	1850–2200	—	8 × 6	—	—	[N]	—	—	21 × 13	Kohn (1961b)
1		ebraeus Linnaeus	Marshall Is.	—	1465	1030–1900	—	5.8–6.3 × 5–5.8	—	—	[27]	—	—	23 × 16	present study
2		ebraeus Linnaeus	Indonesia	160				8 × 7	—	—	[27]	—	—		present study
1	3488	eburneus Hwass	Marshall Is.	150	4700	—	99 000	12–13 × 12.5–14	—	—	[28]	—	—	38 × 24	present study

Appendix Table 1 (*cont.*)

New record no.	Specimen no.	Species	Locality	Egg diam (μm) $\bar{x} \pm$ SD	Eggs per capsule $\bar{x} \pm$ SD	Range	Av. no. eggs per mass	Capsule size (mm) ht × max breadth	Prehatch period (days)	Hatchling size (μm)	Minimum pelagic period (days)	Settling size (SL, mm)	Adult size (mm)	References
2	8485–86	*eburneus* Hwass	Australia	211 ± 11	1339 ± 137	1220–1530	140 000	11–12 × 8.5–10.5	≥13	268	[22]	—	51 × 30; 54 × 35	present study
1	6367	*episcopus* auctt.	Indonesia	400 ± 6	712	693–730	46 000	16–20 × 12–13	>10	—	[6]	—	85 × 34	present study
2	FEP203	*episcopus* auctt.	Palau	275 ± 9	(2148)	1960–2336	—	18 × 12	13	582 ± 16	17	1.45	56 × 35	present study
3	FEP251	*episcopus* auctt.	Palau	277 ± 5	—	—	—	—	—	—	—	—	66 × 27	present study
		figulinus Linnaeus	Sri Lanka	200	—	5700–8600	235 000	18 × 11–12	—	—	[23]	—	57 × 35	Kohn (1960,1961b)
	4029	*figulinus* Reeve	Sri Lanka	—	7000	5800–8200	>208 000	19 × 13	—	—	—	—	53 × 32; 45 × 29	Kohn (1960,1961b)
1	JEN45	*figulinus* Linnaeus	Philippines	193 ± 11	3736 ± 733	3140–4740	228 000	16–17.5 × 10.5	—	—	[24]	—		present study
2	JEN49	*figulinus* Linnaeus	Philippines	209 ± 47	3362 ± 63	3270–3410	104 000	18–21.5 × 10–12.5	—	—	[23]	—	61 × 38	present study
	C158481	*figulinus* Linnaeus	N. Queensland	221 ± 9	—	—	—	25.5–28 × 13–16	—	—	[21]	—	69–76 × 43–46	present study
		flavidus Lamarck	Hawaii	180	—	—	—	14 × 11	15	317	23	1.23	42	Perron (1981a,b)
1	3926	*flavidus* Lamarck	Marshall Is.	175	2400	—	62 000	10–11.5 × 8–9	—	≥285	[25]	—	47 × 29	present study
2	FEP54	*flavidus* Lamarck	Palau	223 ± 3	(2194)	1988–2400	—	10–11 × 10–11	—	—	[21]	—	38 × 23	present study
3	BAW	*flavidus* Lamarck	Cook Is.	[227]	2509 ± 83	2417–2577	—	11.5 × 9.5–10	—	—	[21]	—	45 × 26	present study
	7382	*frigidus* Reeve	India	190 ± 6	3688	3129–4206	122 000	14–17 × 10–11.5	—	—	[24]	—	54 × 34	Kohn (1978)
1	6864	*frigidus* Reeve	Indonesia	212 ± 9	—	—	—	8.5 × 7.5	—	—	[22]	—	33 × 20	present study
2	7141–7147	*frigidus* Reeve	Indonesia	194 ± 9	1567	—	—	6.5 × 7	—	—	[24]	—		present study
3	7150	*frigidus* Reeve	Indonesia	191 ± 3	—	—	—	8–9 × 6	—	—	[24]	—	30 × 17	present study
1	JEN13	*furvus* Reeve	Philippines	—	9.5 ± 2.6	6–12	450	7–8 × 4–5	—	≥900	—	—	39 × 20	present study
2	JEN59	*furvus* Reeve	Philippines	634 ± 20	19 ± 4	14–25	—	7.5–8.5 × 5–5.5	—	—	[N]	—	56 × 29	present study
3	JEN60	*furvus* Reeve	Philippines	697 ± 19	24 ± 4	22–29	3372	9–10 × 6.5	—	—	[N]	—		present study
	5332	*geographus* Linnaeus	Seychelles	190	16 100	14 500–17 800	870 000	26–28 × 18–21	—	240–265	[24]	—	100 × 49	Kohn (1961b)
		geographus Linnaeus	Philippines	—	—	—	—	—	20	—	—	—	—	Cruz et al. (1978)
	4066	*glans* Hwass	Sri Lanka	440	48 ± 8	37–53	1600	7–7.5 × 6–6.5	—	880	[3]	—	24 × 11	Kohn (1961b)
	FEP217	*glans* Hwass	Palau	341 ± 9	(233)	230–236	—	10 × 6–7	14	696	9	1.32	25 × 13	present study
	1983	*imperialis* Linnaeus	Hawaii	225	5900	—	>35 000	18–20 × 12–13	—	—	[21]	—	79 × 23	Kohn (1961a)
	5504	*imperialis* Linnaeus	Seychelles	220	3300	2300–4300	>188 000	18–19 × 11–12	>14	>340	[22]	—	74 × 38	Kohn (1961b)
	FEP–	*imperialis* Linnaeus	Hawaii	225	—	—	—	—	—	370	[21]	1.65	60	Perron (1981); Taylor (1975)
	FEP83	*imperialis* Linnaeus	Palau	265 ± 6	2950	1210–1490	56 000	13 × 9	—	392	[18]	—	60 × 35	present study
		leopardus (Röding)	Hawaii	—	—	—	—	49–58 × 34–37	—	360	—	—		Kohn (1961a)
		leopardus (Röding)	Hawaii	230	—	—	—	45 × 30	16	335	[21]	—	140	Perron (1981a,b)
	5035	*leopardus* (Röding)	Seychelles	225	11 100	10 000–12 800	>744 000	28–31 × 20–22	—	—	[21]	—	80 × 49	Kohn (1961b)
	5040	*leopardus* (Röding)	Seychelles	—	8850	7900–9800	142 000	34–35 × 18–20	—	275–315	[21]	—	108 × 60	Kohn (1961b)
	5507	*leopardus* (Röding)	Seychelles	225	12 300	—	—	33–34 × 23	14	—	—	—	96 × 62	Kohn (1961b)

#	ID	Species	Locality									Reference
	4981	leopardus (Röding)	Seychelles	220	15 600	15 200–16 000	140 000	34–36 × 21–22	—	—	[22]	115 × 65	Kohn (1961b)
	5311	leopardus (Röding)	Seychelles	200	12 550	—	150 000	30–32 × 18–20	—	275–290	[23]	101 × 63	Kohn (1961b)
1		leopardus (Röding)	Marshall Is.	210	—	—	—	45–48 × 24–27	—	300	[22]	—	present study
2		leopardus (Röding)	Marshall Is.	—	—	—	—	47–55 × 28–33	—	—	—	—	present study
3	FEP403	leopardus (Röding)	Palau	204 ± 4	21 400	20 000–22 750	—	34 × 25	—	—	[23]	—	present study
1	7279	litteratus Linnaeus	Indonesia	222 ± 4	9900	9500–10 300	—	27 × 19	—	—	[21]	109 × 65	present study
2	FEP313	litteratus Linnaeus	Palau	201 ± 3	14 000	13 700–14 250	—	28 × 17	—	326	[23]	83 × 46	present study
3		litteratus Linnaeus	Guam	216 ± 8	—	—	—	23 × 16.5	—	—	[22]	—	present study
	1264	lividus Hwass	Hawaii	—	—	—	—	9.5–12 × 9.5–12	—	235–260	—	45 × 26(?)	Kohn (1961a)
	FEP–	lividus Hwass	Hawaii	150	—	—	—	12 × 10	13.5–14.5	250	50	36	Perron (1981a,b)
	4034	lividus Hwass	Sri Lanka	140 × 125	2500	1900–3300	52 000	7–8 × 6.5–7	—	—	—	28 × 17	Kohn (1961b)
	4054	lividus Hwass	Sri Lanka	140	—	—	—	7–8 × 6.5–7	—	—	[28]	33 × 19	Kohn (1961b)
	4053	lividus Hwass	Sri Lanka	142	—	—	—	6–7 × 6–6.5	—	—	[28]	24 × 14	Kohn (1961b)
	4085	lividus Hwass	Maldive Is.	140	—	—	—	7.5 × 6.5	—	—	[28]	31 × 18	Kohn (1961b)
	5686	lividus Hwass	Seychelles	—	—	—	—	15.5–16.5 × 11–11.5	—	≥ 200	—	44 × 26	Kohn (1961b)
1	6151	lividus Hwass	Vietnam	150 ± 9	3200	1800–4700	—	10.5 × 10.5	—	—	[28]	41 × 24	present study
2	6254	lividus Hwass	Indonesia	134 ± 8	3160	2882–3666	139 000	9–10 × 8.5–9.5	—	> 163	[29]	39 × 22	present study
3	7258	lividus Hwass	Indonesia	135 ± 5	9300	9040–9650	—	15 × 12	—	—	[29]	66 × 40	present study
4		lividus Hwass	Indonesia	140 ± 6	7110	5900–8300	—	11–11.5 × 11	—	—	[28]	44	present study
5	JEN12	lividus Hwass	Philippines	150 ± 4	5710	5303–6092	—	12–13 × 10–10.5	—	—	[28]	52 × 28	present study
6a	JEN35A	lividus Hwass	Philippines	151 ± 6	6180	5815–6580	80 000	12–14.5 × 11–13	—	—	[28]	?	present study
6b	JEN35B	lividus Hwass	Philippines	144 ± 6	—	—	—	10.5–11 × 11	—	—	[28]	35 × 20	present study
8	BRAR69	lividus Hwass	Cook Is.	150	3130	2921–3269	> 69 000	9.5–10.5 × 8–9	—	—	[28]	35 × 21	present study
1	6238	magus Linnaeus	Indonesia	566	45 ± 11	27–57	3600	11–12 × 8.5–9	> 15	> 700	[N]	39 × 19	present study
2	6240	magus Linnaeus	Indonesia	554 ± 16.5	38 ± 7	31–44	—	10–11 × 6–8	—	—	[N]	35 × 18	present study
3	3284	magus Linnaeus	Ponape	—	167	139–196	5000	12 × 9.5	—	—	[28]	52 × 23	present study
4	JEN8	magus Linnaeus	Philippines	533	77 ± 12	5303–6092	4400	12–13 × 9–12	—	≥ 1100	[N]	40 × 19	present study
5	JEN10	magus Linnaeus	Philippines	528	47 ± 3	—	—	10 × 9	—	1110–1220	[N]	37 × 18	present study
6	JEN21	magus Linnaeus	Philippines	543 ± 9	64 ± 12	50–74	2500	10–10.5 × 8–8.5	—	≥ 900	[N]	45 × 21 (f); 43 × 20 (f)	present study
7	JEN34	magus Linnaeus	Philippines	561	53 ± 23	—	—	11.5–13 × 10–12	—	≥ 850	[N]	47 × 22	present study
8	JEN46	magus Linnaeus	Philippines	537	92 ± 24	—	1200	12–13 × 9–10.5	—	—	[N]	50 × 24	present study
9	JEN53	magus Linnaeus	Philippines	566	73 ± 18	—	3200	11–13 × 10–11	—	—	[N]	—	present study
10	JEN54	magus Linnaeus	Philippines	512	98 ± 39	—	4200	12–14 × 8–14	—	—	[N]	44 × 21 (f)	present study
11	JEN55	magus Linnaeus	Philippines	499 ± 28	106 ± 12	—	4800	12–13 × 10–12	—	—	[N]	46 × 22; 48 × 22	present study
12	JEN58	magus Linnaeus	Philippines	580	40 ± 31	—	800	11.5–13 × 9–10	—	—	[N]	44 × 22 (f); 50 × 24 (f)	present study
1	3347	marmoreus Linnaeus	Ponape	429	294 ± 124	—	12 000	13.5–16 × 10–12	—	[≥ 550]	[3]	60 × 36	present study

Appendix Table 1 (*cont.*)

New record no.	Specimen no.	Species	Locality	Egg diam (μm) $\bar{x} \pm$ SD	Eggs per capsule $\bar{x} \pm$ SD	Range	Av. no. eggs per mass	Capsule size (mm) ht × max breadth	Prehatch period (days)	Hatchling size (μm)	Minimum pelagic period (days)	Settling size (SL, mm)	Adult size (mm)	References
2	3600	*marmoreus* Linnaeus	Marshall Is.	—	—	—	—	4.5–15.5 × 10–11.5	—	—	—	—	70 × 42	present study
	3610	*marmoreus* Linnaeus	Marshall Is.	450	192 ± 35	140–238	24 600	11–12 × 7–9	—	≥820	[2]	—	54 × 30	present study
3	FEP25	*marmoreus* Linnaeus	Palau	400 ± 7	568	520–616	—	18–19 × 12–14	—	840	8	1.43	93 × 50	present study
4	FEP101	*marmoreus* Linnaeus	Palau	403 ± 6	1858 ± 159	1655–2061	115 200	23–26 × 14–16	16	844	[6]	—	95 × 53	present study
5	JEN16	*marmoreus* Linnaeus	Philippines	409 ± 9	373 ± 96	266–452	—	13.5–14.5 × 9–11	—	—	[5]	—	65 × 35	present study
6	JEN19	*marmoreus* Linnaeus	Philippines	409 ± 24	538	—	52 000	15.5 × 13.5	—	—	[5]	—	72 × 41	present study
7	3444	*miles* Linnaeus	Ponape	228 ± 8	1398	—	9800	19–21 × 10–12	—	—	[21]	—	55 × 34	present study
	4464	*miliaris* Hwass	Maldive Is.	158	982	931–1032	—	5.5 × 6	—	—	[27]	—	21 × 13	Kohn (1961b)
	6031	*miliaris* Hwass	Thailand	202 ± 10	1155	840–1369	≥18 500	9–10.5 × 7–8.5	—	300	[23]	—	32 × 20	present study
	5723	*moreleti* Crosse	Seychelles	150	2350	—	>40 000	10–11 × 8	—	—	[28]	—	38 × 19	Kohn (1961b)
		obscurus Sowerby	Hawaii	147	—	—	—	—	12–13	260	[28]	2.10	27	Perron (1981a)
1	10 247	*omaria* Hwass	Fiji	342	526 ± 42	485–569	13 700	12–13.5 × 8–9.5	—	—	[11]	—	45 × 21	present study
2		*omaria* Hwass	Fiji	—	685	640–730	—	12–13 × 8–9?	—	≥640	—	—	68 × 31	present study
3	FEP61	*omaria* Hwass	Palau	340 ± 5	—	604–912	—	13–16 × 12–15	14	690	12	1.47	63 × 28	present study
4	FEP210	*omaria* Hwass	Palau	331 ± 6	—	—	—	—	14	—	12	—	56 × 24	present study
		pennaceus Born	Hawaii	470	<50	40–150	<1700	11 × 8	16	>1250	N	>1.25	—	Ostergaard (1950)
		pennaceus Born	Hawaii	—	—	—	—	—	—	—	N	—	—	Ostergaard (1950)
	156	*pennaceus* Born	Hawaii	—	86	—	3250	8–8.5 × 7.5–8	—	—	—	—	33 × 19	Kohn (1961a)
	401	*pennaceus* Born	Hawaii	—	—	—	—	—	—	—	—	—	37 × 20	Kohn (1961a)
	402	*pennaceus* Born	Hawaii	—	—	—	—	—	—	—	—	—	37	Kohn (1961a)
	403	*pennaceus* Born	Hawaii	—	—	—	—	—	—	—	—	—	34 × 19	Kohn (1961a)
	404	*pennaceus* Born	Hawaii	490	—	—	—	—	—	—	—	—	33 × 19	Kohn (1961a)
	1810	*pennaceus* Born	Hawaii	—	—	—	—	9.5–11 × 7.5–9	—	—	—	—	[62 × 37 (m)]	Kohn (1961a)
	1811	*pennaceus* Born	Hawaii	—	71	—	2400	9–10.5 × 7–8	25–26?	1200–1300	N	1.2–1.3	35 × 21	Kohn (1961a)
	1962	*pennaceus* Born	Hawaii	—	126	—	8600	12–13 × 9–10	—	1100	N	1.1	—	Kohn (1961a)
		pennaceus Born	Hawaii	500	—	—	—	9–11 × 8	—	—	—	—	39 × 22	Kohn (1961a)
		pennaceus Born	Hawaii	—	—	—	—	12 × 10	24–26	1250	N	1.25	60	Perron (1981a,b)
	4478	*pennaceus* Born	Maldive Is.	390	566	483–662	≥32 800	14–17 × 10–11	—	—	[7]	—	60 × 35	Kohn (1961b)
	7351	*pennaceus* Born	India	628 ± 17	112 ± 3	109–114	3360	15.5–16 × 8.5–9.5	—	≥1140	[N]	≥1.14	53 × 27	present study
	7352	*pennaceus* Born	India	654 ± 12	36	32–40	—	9 × 6	—	—	[N]	—	36 × 18	present study

6318	*pennaceus* Born	Indonesia	407 ± 7	361	322–400	12–14.5 × 8.5–10	—	≥560	[6]	—	50 × 28	present study
	pertusus Hwass	Hawaii	132	—	—	—	11	240	[29]	2.15	27	Perron (1981a); Taylor (1975)
JEN22	*planorbis* Born	Philippines	214 ± 3	—	—	18–19 × 14–15	—	—	[22]	—	64 × 33	present study
C111639	*pulicarius* Hwass	Queensland	145	>1500	—	13–14 × 16–17.5	—	—	[28]	—	—	Huish (1978)
FEP	*pulicarius* Hwass	Hawaii	150	—	—	—	12	280	[27]	1.25	87	Perron (1981)
3489	*pulicarius* Hwass	Marshall Is.	[175]	3350	—	8–9 × 10–11	—	—	[26]	—	27 × 16	present study
1284	*quercinus* [Lightfoot]	Hawaii	215	9700	70 000	22–23 × 20–21	15–16	290	8	—	91 × 56	Kohn (1961a)
FEP	*quercinus* [Lightfoot]	Hawaii	190	—	388 000	—	15.5–16.5	280	30	1.33	130	Perron (1981a,b)
JEN6	*quercinus* [Lightfoot]	Philippines	165	3138	—	20 × 18	—	—	[26]	—	46 × 29	Perron (1981a,b)
JEN9	*quercinus* [Lightfoot]	Philippines	163	3659	2825–3173 147 000?	11.5–12 × 8.5	—	—	[26]	—	59 × 37; 64 × 40	present study
JEN17	*quercinus* [Lightfoot]	Philippines	172 ± 4	1918	3514–3735 893 000	12.5–15 × 8–11.5	—	—	[26]	—	43 × 25	present study
JEN38	*quercinus* [Lightfoot]	Philippines	177 ± 4	883	1689–2148 173 000	9–10 × 6–10.5	—	—	[25]	—	43 × 24	present study
	rattus Hwass	Hawaii	125	2000	826–925 142 000	7–7.5 × 5–6.5	—	—	[30]	—	—	Ostergaard (1950)
	rattus Hwass	Hawaii	[152]	—	— 44 000	—	—	—	[30]	—	—	Ostergaard (1950)
	rattus Hwass	Hawaii	125	—	—	11–15 × 10–14	—	240	[27]	—	37	Kohn (1961a)
	rattus Hwass	Hawaii	125	—	—	12 × 9	11–12	—	[30]	1.35	—	Perron (1981a,b); Taylor (1975)
4010	*rattus* Hwass	Sri Lanka	125	4250	3900–4600 >30 000	9–10.5 × 7	—	—	[30]	—	30 × 18	Kohn (1961b)
5327	*rattus* Hwass	Seychelles	175	6300	4900–7500 >57 000	10–11 × 8–8.5	—	—	[25]	—	27 × 17	Kohn (1961b)
5704	*rattus* Hwass	Seychelles	120	3900	3800–4000 >100 000	8 × 6–7	—	200	[30]	—	27 × 18	Kohn (1961b)
FEP365	*rattus* Hwass	Palau	124 ± 3	5300	4496–6096	11–12 × 9–10	—	—	[19]	—	33 × 20	present study
6157	*retifer* Menke	Indonesia	253 ± 6	239	233–245 23 000	9 × 6–7	—	220	[29]	—	34 × 18	present study
	sponsalis nanus Sowerby	Hawaii	135	—	—	—	11	—	[30]	1.45	19	Perron (1981a); Taylor (1975)
	sponsalis nanus Sowerby	Aldabra	125	1157 ± 100	1056–1270 30 000	4.5–5.5 × 2.5–2.7	—	≥200	[30]	—	16 × 10	present study (coll. J. D. Taylor)
JEN1	*stercusmuscarum* Linnaeus	Philippines	274 ± 9	2734	2699–2787 170 000	7.5–19.5 × 16–18.5	—	—	[17]	—	62 × 32	present study
JEN41	*stercusmuscarum* Linnaeus	Philippines	235 ± 4	1748	1655–1840	15 × 12–14	—	—	[20]	—	52 × 25	present study
EN47	*stercusmuscarum* Linnaeus	Philippines	291 ± 10	1633 ± 85	1561–1727 20 000	16–16.5 × 13–14.5	—	—	[15]	—	49 × 25	present study
FEP115	*stercusmuscarum* Linnaeus	Palau	238 ± 3	1714	1520–1908	18 × 16	—	440	[20]	—	56 × 27	present study
JEN11	*stramineus* Lamarck	Philippines	537	41 ± 11	31–56 2800	9–11 × 6.5–8.5	—	—	[N]	—	29 × 14; 32 × 15	present study
FEP347	*striatellus* Link	Palau	193 ± 5	1836	1800–1872	14–15 × 10	11	290	[24]	—	38 × 20	present study
FEP–	*striatus* Linnaeus	Hawaii	250	—	—	—	16	490	20	1.52	90	Perron (1981a)
1	*striatus* Linnaeus	Australia	267 ± 3	6386	6206–6566 ≥51 000	30 × 25	—	—	[18]	—	89 × 45	present study
2	*striatus* Linnaeus	Australia	262 ± 9	2676?	—	26 × 24	14	—	[18]	—	—	present study
3	*striatus* Linnaeus	Guam	—	—	—	20–22 × 17–20	—	—	—	—	—	present study

Appendix Table 1 (*cont.*)

New record no.	Specimen no.	Species	Locality	Egg diam (μm) $\bar{x} \pm$ SD	Eggs per capsule $\bar{x} \pm$ SD	Range	Av. no. eggs per mass	Capsule size (mm) ht × max breadth	Prehatch period (days)	Hatchling size (μm)	Minimum pelagic period (days)	Settling size (SL, mm)	Adult size (mm)	References
4	JEN29	*strriatus* Linnaeus	Philippines	255	2175	2000–2350	—	16 × 14	—	—	[19]	—	78 × 40	present study
5	JEN40	*strriatus* Linnaeus	Philippines	—	—	—	—	19–20 × 14–17	—	—	—	—	65 × 30	present study
6	FEP348	*strriatus* Linnaeus	Palau	235 ± 3	1817	1784–1850	—	16–17 × 16–19	—	400	[20]	—	73 × 35	present study
	FEP107	*strriolatus* Kiener	Palau	224 ± 5	1202	1171–1232	—	10 × 9	13	440	[21]	—	32 × 18	present study
	JEN52	*suratensis* Hwass	Philippines	207 ± 3	4435	4320–4550	412 000	16–17 × 14–15	—	—	[23]	—	66 × 38 (f); 54 × 35	present study
	C111633	*tessulatus* Born	Persian Gulf	250	246	205–320	—	17–22 × 18–21	—	750	[19]	—	—	Thorson (1940)
	(FEP)—	*textile* Linnaeus	Queensland	230	1300	—	—	31–33 × 21–26	—	—	[21]	—	—	Huish (1978)
		textile Linnaeus	Hawaii	260	—	—	—	21 × 17	17	530	16	1.51	80	Perron (1980, 1981*a*)
	JEN48	*textile* Linnaeus	Philippines	259 ± 6	1038	975–1100	49 000	12 × 8	—	—	[18]	—	56 × 28	present study
	FEP366	*textile* Linnaeus	Palau	253 ± 3	4040	3920–4160	—	18–20 × 13–17	—	450	[19]	—	79 × 37	present study
1	5925	*textile* Linnaeus	Singapore	280	1400	—	78 000	18 × 15	—	—	[17]	—	—	present study
2	6126	*textile* Linnaeus	Thailand	249 ± 6	—	—	—	12.5–13.5 × 10–11.5	—	—	[19]	—	57 × 23	present study
3	6129	*textile* Linnaeus	Thailand	249	2556	2367–2745	143 000	15–16 × 12–14	11	—	[19]	—	62 × 30	present study
4	6413	*textile* Linnaeus	Indonesia	—	—	—	—	19.5–21 × 19	—	440 × 490	—	—	90 × 46	present study
5	JEN20	*textile* Linnaeus	Philippines	—	2870	2438–3073	158 000	16.5–19.5 × 13.5–16	—	≥419	—	—	75 × 33 (f); 73 × 35 (f)	present study
6	JEN27	*textile* Linnaeus	Philippines	—	2152	2068–2301	≥17 000	18–19 × 12–14	—	≥403	—	—	65 × 31	present study
7	JEN28	*textile* Linnaeus	Philippines	248 ± 6	2112	1856–2242	91 000	17.5–19 × 14–16.5	—	—	[19]	—	64 × 32	present study
8	JEN30	*textile* Linnaeus	Philippines	254 ± 11	1234	996–1370	77 000	13–15 × 11.5–13.5	—	—	[19]	—	56 × 24	present study
9	JEN39	*textile* Linnaeus	Fiji	256	1525	1242–1734	59 500	15–16 × 9–13	—	—	[18]	—	54 × 24	present study
10	BFIJ326	*textile* Linnaeus	Philippines	299 ± 4	1894	1862–1937	58 700	17–18.5 × 14–15	—	—	[15]	—	—	present study
1	JEN24	*thalassiarchus* Sowerby	Philippines	602 ± 17	38	28–48	2800	8–9 × 5–6	—	—	[N]	—	56 × 28	present study
2	JEN—	*thalassiarchus* Sowerby	Philippines	—	45	44–47	4100	9.5 × 6–6.5	—	—	—	—	77 × 38	present study
3	JEN25	*thalassiarchus* Sowerby	Philippines	—	23	22–25	2200	7.5–8.5 × 5.5–6	—	≥805	—	—	?	present study
	5874	*varius* Linnaeus	Seychelles	160	1900	1800–2100	63 000	8–10 × 6–?	—	—	[27]	—	34 × 19	Kohn (1961*b*)
	5875	*varius* Linnaeus	Seychelles	—	2000	1700–2700	81 000	9.5–10.5 × 6–7	—	—	—	—	38 × 21	Kohn (1961*b*)
	5876	*varius* Linnaeus	Seychelles	—	—	—	—	9–11 × 7–8	—	≥250	—	—	38 × 22	Kohn (1961*b*)
1	JEN31	*varius* Linnaeus	Philippines	195	2370	2250–2450	—	11.5–12.5 × 8.5–10	—	≥265	[24]	—	42 × 20	present study
2	FEP242	*varius* Linnaeus	Palau	167 ± 3	2000	1750–2300	—	13 × 9	—	—	[26]	—	44 × 20	present study
		vexillum Gmelin	Hawaii	140	—	—	—	20 × 13	≥12	250	[28]	—	—	Ostergaard (1950)
	FEP	*vexillum* Gmelin	Hawaii	140	—	—	—	20 × 16	11.5–12.5	250	[28]	2.00	85	Perron (1981*a*,*b*); Taylor (1975)

	vexillum Gmelin	Seychelles	—	—	—	—	—	—	—	~90	Kohn (1961b)
5518	*vexillum* Gmelin	Seychelles	140	43 500	34 500–53 500	>1 500 000	27.5–31 × 17.5–21.5	—	—	[28]	82 × 50	Kohn (1961b)
1	*victoriae* Reeve	W. Australia	621 ± 21	55 ± 22	25–75	7200	11–14 × 8	—	—	[N]	66 × 23–56 × 29	present paper
4008	*virgo* Linnaeus	Sri Lanka	170	17 000	16 500–18 000	>700 000	30–32 × 20–22	—	—	[26]	104 × 58	Kohn (1961b)
5053–6A	*virgo* Linnaeus	Seychelles	—	5800	—	—	20 × 13	—	—	[25]	66 × 39	Kohn (1961b)
5053–6B	*virgo* Linnaeus	Seychelles	185	—	—	—	12–13 × 12–13	—	—	[23]	58 × 37	Kohn (1961b)
5053–6C	*virgo* Linnaeus	Seychelles	212	5800	3900–7400	220 000	17–19 × 13–15	—	—	[23]	64 × 37	Kohn (1961b)
5058–9	*virgo* Linnaeus	Seychelles	200	—	—	—	15–17.5 × 13–14	—	—	[23]	62 × 35	Kohn (1961b)
JEN36	*virgo* Linnaeus	Philippines	206 ± 6	—	—	—	29 × 22	—	—	[23]	105 × 55	present study
1 3944	*virgo* Linnaeus	Marshall Is.	224 ± 6	—	—	—	15 × 9	—	—	[21]	57 × 31	present study
2 2020	*vitulinus* Hwass	Hawaii	225	—	—	400	23 × 16–17	14–16	—	[21]	67 × 38	Kohn (1961a)
7034	*vitulinus* Hwass	Hawaii	250	—	—	—	—	—	—	[19]	65 × 37	Kohn (1961a)

d = days
(f) = female
(m) = male
N = none
SL = shell length
[ ] = estimate

Appendix Table 2 Summary of data on reproduction and development of *Conus* outside the Indo-Pacific region

Species	Locality	Egg diam (μm) $\bar{x} \pm SD$	Eggs per capsule $\bar{x} \pm SD$	Range	Av. no. eggs per mass	Capsule size (mm) ht × max breadth	EMT	Prehatch period (days)	Hatchling size (μm)	Minimum pelagic period (days)	Settling size (SL, mm)	Adult size (mm)	References
Western Atlantic													
cedonulli Linnaeus	St Vincent	[599]	38	—	—	11.7 × 5.3	—	—	700–800	0	≥1.2–1.4	—	Vink & Cosel (1985)
centurio Born	Golfo Triste, Venezuela	275	1752	1731–1772	56 050	13 × 8	I	—	320	+	—	—	Penchaszadeh (1984)
curralensis Rolán	Curaçao & Bonaire	[555]	—	—	—	—	—	—	700	0	—	—	Vink & Cosel (1985)
ermineus Born	Chengue, Colombia	[233]	500	—	—	20 × 14	I	~10	330	+	—	—	Bandel (1975b, 1976)
ermineus Born	Golfo Triste, Venezuela	[203]	2373	2221–2420	—	19 × 15.5	II	—	295	+	—	—	Penchaszadeh (1984)
mappa [Lightfoot]	Santa Marta to W. Venezuela	[555]	—	—	—	—	—	—	700	0	—	—	Vink & Cosel (1985)
mus Hwass	Chengue, Colombia	[242]	600	500–700	10 200	7 × 5	I	9–12	340	+	—	—	Bandel (1975b, 1976)
mus Hwass	Bermuda	[225]	—	—	—	14	—	—	≥320	—	—	—	Lebour (1945)
mus Hwass	Barbados	120	—	—	—	5	—	10	240	+	—	—	Bandel (1975b)
pseudaurantius Vink & Cosel	Grenada	[555]	—	—	—	—	—	—	700	0	—	—	Vink & Cosel (1985)
puncticulatus Hwass	Santa Marta, Colombia	[660]	15	9–21	225	6 × 3.5	I	18	820	0	—	—	Bandel (1975b, 1976)
regius Gmelin	Long Reef, Florida	—	—	—	—	16.6 × 11.5	I	—	—	—	—	—	D'Asaro (1970)
regius Gmelin	Chengue, Colombia	[242]	—	100–200	6300	11 × 7	I	10	340	+	—	—	Bandel (1975b; 1976)
spurius Gmelin	Biscayne Bay, Florida	—	59	54–67	170–5100	12.0 × 8.0	I	—	—	—	—	—	D'Asaro (1970)
spurius Gmelin	Gulf of Uruba, Colombia	[1077]	30	20–40	440	12 × 7	II	~12	1300	0	2.1	—	Bandel (1975b; 1976)
spurius Gmelin	Golfo Triste, Venezuela	690	46	23–83	1165	11–13 × 7–9	II	—	880	—	1.3	—	Penchaszadeh (1984)
Eastern Atlantic													
antoniomonteiroi Rolán	Cape Verde Is.	—	—	6–10	—	4–6	—	—	1430	[0]	1.4	—	Rolán (1990)
ateralbus Kiener	Cape Verde Is.	—	16.5	15–18	495	9–11	I	—	—	—	—	—	Rolán (1985, 1986a)
boavistensis Rolán & Fernandez	Cape Verde Is.	560	3	—	—	2	I	—	—	—	—	—	Rolán (1990)
cuneolus Reeve	Cape Verde Is.	—	9.5	7–12	124	4–8	I	45	480	0	—	—	Rolán (1985, 1986a)
curralensis Rolán	Cape Verde Is.	[477]	—	—	—	4	I	—	480–760?	—	—	14–23	Rolán (1986a)
damottae galeo Rolán	Cape Verde Is.	—	—	8–10	—	3–4	—	—	1100	[0]	1.1	—	Rolán (1990)
decoratus Röckel et al.	Cape Verde Is.	—	—	—	—	—	I	—	—	0	—	—	Rolán (1985, 1986a)
derrubado Rolán & Fernandez	Cape Verde Is.	[434]	—	20–25	—	8–10	I	—	1280–1340	[0]	1.28–1.34	—	Rolán (1990)
diminutus Trovão & Rolán	Cape Verde Is.	—	4.5	4–5	—	2–3	I	—	720	—	—	8–19	Trovão & Rolán (1986)
ermineus Born	Cape Verde Is.	180	3250	—	—	10–19 × 7–13	II	—	330	+	—	—	Rolán (1986b)
felitae Rolán	Cape Verde Is.	—	—	4–7	—	3	—	—	1080–1830	[0]	1.1	—	Rolán (1990)
fontonae Rolán & Trovão	Cape Verde Is.	—	—	6–10	—	3–5	—	—	1300–1340	[0]	1.3	—	Rolán (1990)
guinaicus Hwass	Canary Islands	[581]	35	30–40	—	10 × 6	II	—	900	0	—	—	Bandel (1975b); Fioroni (1966)
infinitus Rolán	Cape Verde Is.	—	—	6–10	—	3–4	—	—	1200	[0]	1.2	—	Rolán (1990)
mercator Linnaeus	Dakar, Senegal	400	32 ± 18	14–59	—	3–7	II	—	>620	—	—	—	Knudsen (1950)
mordeirae Rolán & Trovão	Cape Verde Is.	—	—	15–25	—	6–8	—	—	1170–1200	[0]	1.2	—	Rolán (1990)
navarroi Rolán	Cape Verde Is.	[425]	7	4–10	—	3–4	I	—	480	—	—	11–19	Rolán (1986a)
regonae Rolán & Trovão	Cape Verde Is.	—	—	6–20	—	4–6	—	—	1080–1150	[0]	1.1	—	Rolán (1990)

Species	Locality	SL										Reference
saragasae Rolán	Cape Verde Is.	[347]	9	6–12	—	5	I	—	460	—	12–22	Rolán (1986a)
serranegrae Rolán	Cape Verde Is.	—	—	4–5	—	3	—	—	—	—	—	Rolán (1990)
teodorae Rolán & Fernandez	Cape Verde Is.	560	12	—	—	5–6	—	—	1300–1400	[0]	1.3–1.4	Rolán (1990)
ventricosus Gmelin	Banyuls-sur-Mer, France	530–570; 550	11	4–21	33–66	6 × 3–4	I	21	900–1100	0	0.9–1.1	Bandel (1975a); Fioroni (1966); Bandel & Wils (1977)
verdensis furnae Rolán	Cape Verde Is.	—	—	6–12	—	4–5	—	—	1400	[0]	1.6	Rolán (1990)
Eastern Pacific												
purpurascens Broderip	Guaymas, Mexico	<185	5650	5400–5900	—	10–15 × 9–11	I	—	≥200	—	—	Nybakken (1970)
Temperate Australia												
anemone Lamarck	—	—	—	—	—	—	—	—	—	0	—	Wilson & Gillett (1971)
anemone Lamarck	Rottnest I., WA	765–780 × 690–708	17	—	—	9–11 × 7–8	I	—	1640	0	1.64	Kohn (1992)
anemone Lamarck	Rottnest I., WA	756–830 × 671–744	16 ± 6	9–20	—	7–8 × 5–6	I	—	≥1590	[0]	≥1.59	36 × 19 Kohn (1992)
anemone Lamarck	Rottnest I., WA	—	22	—	308	7–8 × 5–6	I	—	—	0	1.61	25 × 14 Kohn (1993)
anemone Lamarck	Barrow I., WA	—	21	—	—	7–9 × 4	I	—	—	[0]	—	Kohn (1993)
dorreensis Péron	Rottnest I., WA	152–155	1014	888–1141	14 200	6 × 6	I	[12–17]	—	—	—	30 × 16 Kohn (1993)
dorreensis Péron	Rottnest I., WA	—	929	—	≥35 300	7 × 5	I	[12–17]	—	—	—	26 × 16 Kohn (1993)
dorreensis Péron	Rottnest I., WA	154	963	949–976	—	7 × 6	I	[12–17]	—	—	—	24 × 15 Kohn (1993)
dorreensis Péron	Rottnest I., WA	150	1422	1126–1585	39 800	6–7 × 5–6	I	[12–17]	—	—	—	Kohn (1993)
dorreensis Péron	Rottnest I., WA	—	1376	1365–1387	—	7–8 × 5–6	I	[12–17]	238	—	—	Kohn (1993)
dorreensis Péron	Rottnest I., WA	154	2118	—	—	10.5 × 10.5	I	[12–17]	238–262	—	—	33 × 18 Kohn (1993)
dorreensis Péron	Rottnest I., WA	152	1464	—	—	5 × 6	I	[12–17]	—	—	—	26 × 15 Kohn (1993)
dorreensis Péron	Rottnest I., WA	—	1454	—	≥10 200	5.5 × 6	I	[12–17]	—	—	—	Kohn (1993)
dorreensis Péron	Rottnest I., WA	—	1963	1763–2117	70 700	7 × 8	I	[12–17]	—	—	—	Kohn (1993)
dorreensis Péron	Rottnest I., WA	152	1630	1504–1757	39 100	7 × 7–8	I	[12–17]	—	—	—	Kohn (1993)
dorreensis Péron	Rottnest I., WA	—	1416	—	17 700	6–7 × 7–8	I	[12–17]	—	—	—	Kohn (1993)
dorreensis Péron	Rottnest I., WA	150	—	—	—	9 × 7	I	[12–17]	—	—	—	Kohn (1993)
dorreensis Péron	Barrow I., WA	160	2462	1048–3648	≥24 600	10 × 11	I	[12–17]	430	0	1.3	43 × 24 Kohn (1993)
klemae Cotton	Rottnest I., WA	279 ± 7.9	1078	—	61 400	12 × 13	I	—	1280	—	—	Kohn (1993)
papilliferus Sowerby	—	1000	—	—	—	—	—	—	—	—	—	Wilson & Gillett (1971)
South Africa												
mozambicus Hwass	—	—	20 ± 2	19–23	—	17 × 6.7	—	≥2.3	—	0	—	Kilburn (1971); Kilburn & Rippey (1982)
scitulus Reeve	—	—	1–5	6–16	—	19 × 8	—	—	—	[0]	—	Kilburn (1971)
tinianus Hwass	—	1000	12 ± 5	—	—	9 × 7	—	2.6	—	[0]	—	Kilburn (1971)

+ = present
[ ] = estimate
EMT = egg mass type
SL = shell length
WA = Western Australia
md = maximum diameter

Single dimensions under capsule size indicate height.

REFERENCES

Adams, H. and Adams, A. (1853). *The genera of recent Mollusca arranged according to their organisation.* van Voorst, London.

Amio, M. (1963). A comparative embryology of marine gastropods, with ecological emphasis. *Jour. Shimonoseki Coll. Fisheries,* **12**: 229–53.

Ankel, W. E. (1929). Über die Bildung der Eikapsel bei *Nassa*-Arten. *Zool. Anz.,* **4**: 219–53.

Bandel, K. (1975*a*). Das Embryonalgehäuse mariner Prosobranchier der Region von Banyuls-sur-Mer. *Vie et Milieu,* **25**: 83–118.

Bandel, K. (1975*b*). Embryonalgehäuse Karibischer Meso- und Neogastropoden (Mollusca). *Abh. Mathem.-Naturw. Klasse, Akad. Wiss. Lit. Mainz,* **1**: 1–133.

Bandel, K. (1976). Spawning, development and ecology of some higher Neogastropoda from the Caribbean Sea of Colombia (South America). *Veliger,* **19**: 176–93.

Bandel, K. and Wils, E. (1977). On *Conus mediterraneus* and *Conus guinaicus*. *Basteria,* **41**: 33–45.

Barkati, S. and Ahmed, M. (1985). Egg capsules and larvae of two species of *Conus* from the coast of Pakistan bordering the northern Arabian Sea. *Pakistan Jour. Zool.,* **17**: 387–92.

Bergh, R. (1895). Beiträge zur Kenntnis der Coniden. *Nova Acta Ksl. Leop.-Carol. Akad. Naturf.,* **65**: 67–214.

Bieler, R. and Hadfield, M. G. (1990). Reproductive biology of the sessile gastropod *Vermicularia spirata* (Cerithioidea: Turritellidae). *Jour. Moll. Stud.,* **56**: 205–19.

Boring, L. (1989). Cell–cell interactions determine the dorsoventral axis in embryos of an equally cleaving opisthobranch mollusc. *Devel. Biol.,* **136**: 239–53.

Bouchet, P. (1989). A review of poecilogony in gastropods. *Jour. Moll. Stud.,* **55**: 67–78.

Brothers, E. B. and Thresher, R. E. (1985). Pelagic duration, dispersal, and the distribution of Indo-Pacific coral-reef fishes. *NOAA Symp. Ser. Undersea Res.,* **3**: 53–69.

Burch, J. B. (1980). A guide to the freshwater shells of the Philippines. *Malacol. Rev.,* **13**: 123–43.

Catterall, C. P. and Poiner, I. R. (1983). Age- and sex-dependent patterns of aggregation in the tropical gastropod *Strombus luhuanus*. *Mar. Biol.,* **77**: 171–82.

Cernohorsky, W. O. (1964). The Conidae of Fiji. *Veliger,* **7**: 61–94.

Christiansen, F. and Fenchel, T. (1979). Evolution of marine invertebrate reproductive patterns. *Theor. Pop. Biol.,* **16**: 267–81.

Congdon, J. D. and Gibbons, J. W. (1987). Morphological constraint on egg size: a challenge to optimal egg size theory? *Proc. Nat. Acad. US,* **84**: 4145–7.

Coomans, H. E. and DeVisser, J. S. (1987). Studies on Conidae (Mollusca: Gastropoda). 10. The holotype and identity of *Conus coffeae* Gmelin. *Veliger,* **29**: 437–41.

Coomans, H. E., Moolenbeek, R. G., and Wils, E. (1979–86). Alphabetical revision of the (sub)species in recent Conidae. *Basteria,* **43**: 9–26; **43**: 81–115; **44**: 17–49; **45**: 3–55; **46**: 3–67; **47**: 67–143; **49**: 145–96; **50**: 93–150.

Cruz, L. J., Corpuz, G., and Olivera, B. M. (1978). Mating, spawning, development and feeding habits of *Conus geographus* in captivity. *Nautilus,* **92**: 150–3.

D'Asaro, C. (1970). Egg capsules of prosobranch mollusks from south Florida and the Bahamas and notes on spawning in the laboratory. *Bull. Mar. Sci.,* **20**: 414–40.

D'Asaro, C. (1988). Micromorphology of neogastropod egg capsules. *Nautilus,* **102**: 134–48.

Dautzenberg, P. (1937). Gastéropodes marins. 3. Famille Conidae. *Mém. Mus. Roy. Hist. Nat. Belg.,* **2**(18): 1–184.

Duarte, C. M. and Alcaraz, M. (1989). To produce many small or a few large eggs: a size-dependent reproductive tactic of fish. *Oecologia,* **80**: 401–4.

Emerson, W. K. (1978). Mollusks with Indo-Pacific faunal affinities in the eastern Pacific Ocean. *Nautilus,* **92**: 91–6.

Emerson, W. K. (1983). New records of prosobranch gastropods from Pacific Panama. *Nautilus,* **97**: 119–23.

Emlet, R. B., McEdward, L. R., and Strathmann, R. R. (1987). Echinoderm larval ecology viewed from the egg. In *Echinoderm Studies* (ed. M. Jangoux and J. M. Lawrence), **2**: 55–136.

Estival, J. (1981). *Cônes de Nouvelle-Calédonie et du Vanuatu.* Société Nouvelle des Editions du Pacifique, Papeete.

Felsenstein, J. (1985). Phylogenies and the comparative method. *Amer. Nat.*, **125**: 1–15.

Fioroni, P. (1966). Zur Morphologie und Embryogenese des Darmtraktes und der transitorischen Organe bei Prosobranchiern (Mollusca, Gastropoda). *Rev. Suisse Zool.*, **73**: 621–876.

Frank, P. W. (1969). Growth rates and longevity of some gastropod mollusks on the coral reef at Heron Island. *Oecologia*, **2**: 232–50.

Fretter, V. (1941). The genital ducts of some British stenoglossan prosobranchs. *Jour. Mar. Biol. Assn. UK*, **25**: 173–211.

Fretter, V. (1984). Prosobranchs. In *The Mollusca*, Vol. 7, *Reproduction* (ed. A. S. Tompa, N. H. Verdonk, and J. A. M. van den Biggelaar), pp. 1–45. Academic Press, Orlando.

Futuyma, D. (1986). *Evolutionary biology*. Sinauer Assoc., Sunderland, Mass.

Gaston, K. J. (1990). Patterns in the geographical ranges of species. *Biol. Rev.*, **65**: 105–29.

Grant, A. (1983). On the evolution of brood protection in marine benthic invertebrates. *Amer. Nat.*, **122**: 549–55.

Hadfield, M. G. (1989). Latitudinal effects on juvenile size and fecundity in *Petaloconchus* (Gastropoda). *Bull. Mar. Sci.*, **45**: 369–76.

Hadfield, M. G. and Miller, S. E. (1987). On developmental pattern of opisthobranchs. *Amer. Malacol. Bull.*, **5**: 197–214.

Hadfield, M. G. and Strathmann, M. F. (1990). Heterostrophic shells and pelagic development in trochoideans: implications for classification, phylogeny and palaeoecology. *Jour. Moll. Stud.*, **56**: 239–56.

Hadfield, M. G. and Switzer-Dunlap, M. (1984). Opisthobranchs. In *The Mollusca*. Vol. 7, *Reproduction* (ed. A. S. Tompa, N. H. Verdonk, and J. A. M. van den Biggelaar), pp. 209–350. Academic Press, Orlando.

Hart, M. W. (1992). Larval feeding performance and egg size evolution in echinoids. *Amer. Zool.*, **32**: 114A.

Hansen, T. (1980). Influence of larval dispersal and geographic distribution on species longevity in neogastropods. *Paleobiology*, **6**: 193–207.

Hermans, C. O. (1979). Egg size and energetics. In *Reproductive ecology of marine invertebrates* (ed. S. E. Stancyk), pp. 1–9. Univ. of South Carolina Press, Columbia, SC.

Highsmith, R. D. (1985). Floating and rafting as potential dispersal mechanisms in brooding invertebrates. *Mar. Ecol. Progr. Ser.*, **25**: 169–79.

Hoagland, K. E. (1986). Patterns of encapsulation and brooding in the Calyptraeidae (Prosobranchia; Mesogastropoda). *Amer. Malacol. Bull.*, **4**: 173–83.

Hoagland, K. E. and Robertson, R. (1988). An assessment of poecilogony in marine invertebrates: phenomenon or fantasy? *Biol. Bull.*, **174**: 109–25.

Hughes, R. N. and Roberts, D. J. (1980). Reproductive effort of winkles (*Littorina* spp.) with contrasted methods of reproduction. *Oecologia*, **47**: 130–6.

Huish, P. J. (1978). Factors influencing the distribution of *Conus* in east Australian waters. B.Sc. thesis, Dept. of Geography, Univ. of Newcastle, NSW, Australia.

Jablonski, D. and Lutz, R. (1980). Larval shell morphology: ecological and paleontological applications. In *Skeletal growth of aquatic organisms* (ed. D. C. Rhoads and R. A. Lutz), pp. 441–53. Hunt Institute for Botanical Documentation, Carnegie-Mellon University, Pittsburgh, Pennsylvania.

Jablonski, D. and Lutz, R. (1983). Larval ecology of marine benthic invertebrates: paleontological implications. *Biol. Rev.*, **58**: 21–89.

Jackson, G. A. and Strathmann, R. R. (1981). Larval mortality from offshore mixing as a link between precompetent and competent periods of development. *Amer. Nat.*, **118**: 16–26.

Jokiel, P. L. (1984). Long distance dispersal of reef corals by rafting. *Coral Reefs*, **3**: 113–16.

Kay, E. A. (1990). Turrid faunas of Pacific Islands. *Malacologia*, **32**: 79–87.

Keen, A. M. (1971). *Sea shells of tropical west America*. Stanford University Press, Stanford, California.

Kempf, S. C. and Hadfield, M. G. (1985). Planktotrophy by the lecithotrophic larvae of a nudibranch, *Phestilla sibogae* (Gastropoda). *Biol. Bull.*, **169**: 119–30.

Kempf, S. C. and Todd, C. (1989). Feeding potential in the lecithotrophic larvae of *Adalaria proxima* and *Tritonia hombergi*, an evolutionary perspective. *Jour. Mar. Biol. Assn. UK*, **69**: 659–82.

Kilburn, R. N. (1971). A revision of the littoral Conidae (Mollusca: Gastropoda) of the Cape Province. *Ann. Natal Mus.*, **21**: 37–54.

Kilburn, R. and Rippey, E. (1982). *Sea shells of southern Africa*. Macmillan, Johannesburg, South Africa.

Knudsen, J. (1950). Egg capsules and development of some marine prosobranchs from tropical West Africa. *Atlantide Report*, **1**: 85–130.

Kohn, A. J. (1960). Ecological notes on *Conus* (Mollusca: Gastropoda) in the Trincomalee region of Ceylon. *Ann. Mag. Nat. Hist.*, Ser. 13, **2**: 309–20.

Kohn, A. J. (1961a). Studies on spawning behavior, egg

masses, and larval development in the gastropod genus *Conus*. I. Observations on nine species in Hawaii. *Pac. Sci.*, **14**: 163–79.

Kohn, A. J. (1961*b*). Studies on spawning behavior, egg masses, and larval development in the gastropod genus *Conus*. I. Observations in the Indian Ocean during the Yale Seychelles Expedition. *Bull. Bingham Oceanogr. Coll.*, **17**: 3–51.

Kohn, A. J. (1978). The Conidae (Mollusca: Gastropoda) of India. *Jour. Nat. Hist.*, **12**: 295–335.

Kohn, A. J. (1981). Abundance, diversity, and resource use in an assemblage of *Conus* species in Enewetak Lagoon. *Pacific Sci.*, **34**: 359–69.

Kohn, A. J. (1983). Feeding biology of Gastropods. In *The Mollusca* (ed. A. S. M. Saleuddin and K. M. Wilbur), Vol. 5, pp. 1–63. Academic Press, New York.

Kohn, A. J. (1985). Evolutionary ecology of *Conus* on Indo-Pacific coral reefs. *Proc. Fifth Int. Coral Reef Cong.*, **4**: 139–44.

Kohn, A. J. (1987). Intertidal ecology of Enewetak Atoll. In *The natural history of Enewetak Atoll*, Vol. 1, The ecosystem: environments, biotas, and processes (ed. D. Devaney, E. S. Reese, B. L. Burch, and P. Helfrich), pp. 139–57. US Department of Energy.

Kohn, A. J. (1989). Natural history and the necessity of the organism. *Amer. Zool.*, **29**: 1095–103.

Kohn, A. J. (1993). Development and early life history of three temperate Australian species of *Conus* (Mollusca: Gastropoda). *Proc. Fifth Int. Mar. Biol. Workshop*. In press.

Kohn, A. J. and Leviten, P. J. (1976). Effect of habitat complexity on population density and species richness in tropical intertidal predatory gastropod assemblages. *Oecologia*, **25**: 199–210.

Kohn, A. J. and Nybakken, J. W. (1975). Ecology of *Conus* on eastern Indian Ocean fringing reefs: diversity of species and resource utilization. *Mar. Biol.*, **29**: 211–34.

Kool, S. P. (1987). Significance of radular characters in reconstruction of thaidid phylogeny (Neogastropoda: Muricacea). *Nautilus*, **101**: 117–32.

Korn, W. (1990). On the taxonomical status of *Conus dictator* Melvill, 1898. *La Conchiglia*, **22**: 22–5.

LaBarbera, M. (1989). Analyzing body size as a factor in ecology and evolution. *Annu. Rev. Ecol. Syst.*, **20**: 97–117.

Lebour, M. V. (1945). The eggs and larvae of some prosobranchs from Bermuda. *Proc. Zool. Soc. Lond.*, **114**: 462–89.

Lessios, H. A. (1990). Adaptation and phylogeny as determinants of egg size in echinoderms from the two sides of the isthmus of Panama. *Amer. Nat.*, **135**: 1–13.

Leviten, P. J. and Kohn, A. J. (1980). Microhabitat resource use, activity patterns, and periodic catastrophe: *Conus* on tropical intertidal reef rock benches. *Ecol. Monogr.*, **50**: 56–75.

Lewis, J. B. (1960). The fauna of rocky shores of Barbados, West Indies. *Can. Jour. Zool.*, **38**: 391–435.

Lima, G. M. and Lutz, R. A. (1990). The relationship of larval shell morphology to mode of development in marine prosobranch gastropods. *Jour. Mar. Biol. Assn. UK*, **70**: 611–37.

Lima, G. M. and Pechenik, J. A. (1985). The influence of temperature on growth rate and length of larval life of the gastropod *Crepidula plana* (L.). *Jour. Exp. Mar. Biol. Ecol.*, **90**: 55–71.

Linnaeus, C. (1758). *Systema naturae per regna tria naturae*, 10th edn. Stockholm.

Marsh, J. A. (1974). *Cone shells of the world*. Jacaranda Press, Milton, Queensland.

McArdle, B. H. (1988). The structural relationship: regression in biology. *Can. Jour. Zool.*, **66**: 2329–39.

McEdward, L. R. (1984). Morphometric and metabolic analysis of the growth and form of an echinopluteus. *Jour. Exp. Mar. Biol. Ecol.*, **82**: 259–87.

McEdward, L. R. (1988). Experimental embryology as a tool for studying the evolution of echinoderm life histories. In *Echinoderm phylogeny and evolutionary biology* (ed. C. R. C. Paul and A. B. Smith), pp. 189–96. Clarendon Press, Oxford.

Menge, B. A. (1975). Brood or broadcast? The adaptive significance of different reproductive strategies in two intertidal sea-stars *Leptasterias hexactis* and *Pisaster ochraceus*. *Mar. Biol.*, **31**: 87–100.

Miller, S. L. (1974). Adaptive design of locomotion and foot form in prosobranch gastropods. *Jour. Exp. Mar. Biol. Ecol.*, **14**: 99–156.

Natarajan, A. V. (1957). Studies on the egg masses and larval development of some prosobranchs from the Gulf of Mannar and the Palk Bay. *Proc. Indian Acad. Sci.*, **46**: 170–228.

Nussbaum, R. A. and Schultz, D. L. (1989). Coevolution of parental care and egg size. *Amer. Nat.*, **133**: 591–603.

Nybakken, J. (1970). Notes on the egg capsules and larval development of *Conus purpurascens* Broderip. *Veliger*, **12**, 480–1.

Nybakken, J. W. and Perron, F. (1988). Ontogenetic change in the radula of *Conus magus* (Gastropoda). *Mar. Biol.*, **98**: 239–42.

Ó Foighil, D. (1989). Planktotrophic larval development is associated with a restricted geographic range in *Lasaea*, a genus of brooding, hermaphroditic bivalves. *Mar. Biol.*, **103**: 349–58.

Olivera, B. M., Rivier, J., Clark, C., Ramilo, C. A., Corpuz, G. P., Abogadie, F. C., Mena, E. E., Woodward, S. R., Hillyard, D. R., and Cruz, L. J. (1990). Diversity of *Conus* neuropeptides. *Science*, **249**: 257–63.

Ostergaard, J. M. (1950). Spawning and development in some Hawaiian marine gastropods. *Pac. Sci.*, **4**: 75–115.

Pawlik, J. H., O'Sullivan, J. B., and Harasewych, M. G. (1988). The egg capsules, embryos, and larvae of *Cancellaria cooperi* (Gastropoda; Cancellariidae). *Nautilus*, **102**: 47–53.

Pechenik, J. A. (1980). Growth and energy balance during the larval lives of three prosobranch gastropods. *Jour. Exp. Mar. Biol. Ecol.*, **44**: 1–28.

Penchaszadeh, P. (1984). Observations on the spawn of three species of *Conus* from the Golfo Triste, Venezuela. *Veliger*, **27**: 14–18.

Perrin, N., Ruedi, M., and Saiah, H. (1987). Why is the cladoceran *Simocephalus vetulus* (Müller) not a 'bang-bang strategist'? *Functional Ecol.*, **1**: 223–8.

Perron, F. E. (1980). Laboratory culture of the larvae of *Conus textile* Linné (Gastropoda: Toxoglossa) in Hawaii, USA. *Jour. Exp. Mar. Biol. Ecol.*, **42**: 27–38.

Perron, F. E. (1981a). Larval biology of six species of the genus *Conus* (Gastropoda: Toxoglossa) in Hawaii, USA. *Mar. Biol.*, **61**: 215–20.

Perron, F. E. (1981b). The partitioning of reproductive energy between ova and protective capsules in marine gastropods of the genus *Conus*. *Amer. Nat.*, **118**: 110–18.

Perron, F. E. (1981c). Larval growth and metamorphosis of *Conus* (Gastropoda: Toxoglossa) in Hawaii. *Pac. Sci.*, **35**: 25–38.

Perron, F. E. (1982). Inter- and intraspecific patterns of reproductive effort in four species of cone shells (*Conus* spp.). *Mar. Biol.*, **68**: 161–7.

Perron, F. E. (1983). Growth, fecundity and mortality of *Conus pennaceus* in Hawaii. *Ecology*, **64**: 53–62.

Perron, F. E. (1986). Life history consequences of differences in developmental mode among gastropods in the genus *Conus*. *Bull. Mar. Sci.*, **39**: 485–97.

Perron, F. E. and Carrier, R. H. (1981). Egg size distributions among closely related marine invertebrate species: are they bimodal or unimodal? *Amer. Nat.*, **118**: 749–55.

Perron, F. E. and Corpuz, G. C. (1982). Costs of parental care in the gastropod *Conus pennaceus*: age specific changes and physical constraints. *Oecologia*, **55**: 319–24.

Perron, F. E. and Kohn, A. J. (1985). Larval dispersal and geographic distribution in coral reef gastropods of the genus *Conus*. *Proc. Fifth Int. Coral Reef Cong., Tahiti*, **4**: 95–100.

Ponder, W. F. and Clark, G. A. (1989). A morphological and electrophoretic examination of '*Hydrobia buccinoides*', a variable brackish-water gastropod from temperate Australia. *Aust. Jour. Zool.*, **36**: 661–89.

Powell, A. W. B. (1942). The New Zealand recent and fossil Mollusca of the family Turridae. *Bull. Auckland Inst. Mus.*, No. 2.

Ramón, M. (1990). Spawning and development of *Calliostoma granulatum* in the Mediterranean Sea. *Jour. Mar. Biol. Assn. UK*, **70**: 321–8.

Rawlings, T. A. (1990). Associations between egg capsule morphology and predation among populations of the marine gastropod, *Nucella emarginata*. *Biol. Bull.*, **179**: 312–25.

Rice, M. E. (1989). Comparative observations of gametes, fertilization, and maturation in sipunculans. In *Reproduction, genetics and distribution of marine organisms* (ed. J. S. Ryland and P. A. Tyler). 23rd Euro. Mar. Biol. Symp., pp. 167–82, Fredensborg, Denmark.

Risbec, J. (1932). Notes sur la ponte et le développement de mollusques gastéropodes de Nouvelle-Caledonie. *Bull. Soc. Zool. France*, **57**: 358–74.

Robertson, R. (1976). Marine prosobranch gastropods: larval studies and systematics. *Thalass. Jugoslav.*, **10**: 213–38.

Robertson, R. (1985). Archaeogastropod biology and the systematics of the genus *Tricolia* (Trochacea: Tricoliidae) in the Indo-West Pacific. *Monogr. Mar. Moll.*, No. 3.

Röckel, D., Korn, W., and Kohn, A. J. (1993). *A manual of Conidae*, Vol. 1. Verlag Christa Hemmen. In press.

Rolán, E. (1985). Aportaciones al conocimiento de los *Conus* de Cabo Verde por las observaciones realizados en acuario. *Thalassas*, **3**: 37–56.

Rolán, E. (1986a). Description de tres nuevas especies del genero *Conus* (Gastropoda) del Archipelago de Cabo Verde. *Publ. Ocas. Soc. Port. Malac.*, **6**: 1–16.

Rolán, E. (1986b). Aportaciones al conocimiento de

Conus ermineus Born, 1778 (Gastropoda: Conidae): Estudio de los estadios juveniles. *Boll. Malacologico*, **22**: 285–92.

Rolán, E. (1990). Descripcion de nuevas especies y subespecies del genero *Conus* (Mollusca, Neogastropoda) para el archipelago de Cabo Verde. *Iberus*, Sup. **2**: 5–70.

Rosen, B. R. (1988). Progress, problems and patterns in the biogeography of reef corals and other tropical marine organisms. *Helgoländer Meeresunters.*, **42**: 269–301.

Rosenblatt, R. H. and Waples, R. S. (1986). A genetic comparison of allopatric populations of shore fish species from the eastern and central Pacific Ocean: Dispersal or vicariance? *Copeia*, **1986**: 275–84.

Rumphius, G. E. (1705). *D'Amboinsche Rariteitkamer*. Amsterdam.

Sastry, A. N. (1979). Pelecypoda (excluding Ostreidae). In *Reproduction of marine invertebrates* (ed. A. C. Giese and J. S. Pearse), **5**: 113–292. Academic Press, New York.

Scheltema, R. S. (1986a). On dispersal and planktonic larvae of benthic invertebrates: an eclectic overview and summary of problems. *Bull. Mar. Sci.*, **39**: 290–322.

Scheltema, R. S. (1986b). Long distance dispersal by planktonic larvae of shoal-water benthic invertebrates among central Pacific islands. *Bull. Mar. Sci.*, **39**: 241–56.

Scheltema, R. S. (1989). Planktonic and non-planktonic development among prosobranch gastropods and its relationship to the geographic range of species. Proc. 23rd Euro. Mar. Biol. Symp.

Sedgewick, R. (1983). *Algorithms*. Addison-Wesley, Reading, Mass.

Shimek, R. L. (1986). The biology of the Northeastern Pacific Turridae. V. Demersal development, synchronous settlement and other aspects of the larval biology of *Oenopota levidensis*. *Int. Jour. Inv. Reprod. Devel.*, **10**: 313–33.

Shine, R. (1978). Propagule size and parental care: the 'safe harbor' hypothesis. *Jour. Theor. Biol.*, **75**: 417–24.

Shine, R. (1989). Alternative models for the evolution of offspring size. *Amer. Nat.*, **134**: 311–17.

Shuto, T. (1974). Larval ecology of prosobranch gastropods and its bearing on biogeography and paleontology. *Lethaia*, **7**: 239–56.

Sibly, R., Calow, P., and Nichols, N. (1985). Are patterns of growth adaptive? *Jour. Theor. Biol.*, **112**: 553–74.

Sinervo, B. (1990). The evolution of maternal investment in lizards: an experimental and comparative analysis of egg size and its effects on offspring performance. *Evolution*, **44**: 279–94.

Sinervo, B., Doughty, P., Huey, R. B., and Zamudio, K. (1992). Allometric engineering: a causal analysis of natural selection on offspring size. *Science*, **258**: 1927–30.

Sinervo, B. and Huey, R. B. (1990). Allometric engineering: an experimental test of the causes of interpopulational differences in performance. *Science*, **248**: 1106–9.

Sinervo, B. and McEdward, L. R. (1988). Developmental consequences of an evolutionary change in egg size: an experimental test. *Evolution*, **42**: 885–99.

Smith, B. J., Black, J. H., and Shepherd, S. A. (1989). Molluscan egg masses and capsules. In *Marine invertebrates of South Australia* (ed. S. A. Shepherd and I. M. Thomas), Vol. 2.

Smith, C. C. and Fretwell, S. D. (1974). The optimal balance between size and number of offspring. *Amer. Nat.*, **108**: 499–506.

Sokal, R. R. and Rohlf, F. J. (1981). *Biometry*, 2nd edn. W. H. Freeman, San Francisco.

Spight, T. M. (1974). Sizes of populations of a marine snail. *Ecology*, **55**: 712–29.

Spight, T. M. (1975). Factors extending gastropod embryonic development and their selective cost. *Oecologia*, **21**: 1–16.

Spight, T. M. (1976). Ecology of hatching size for marine snails. *Oecologia*, **24**: 283–94.

Spight, T. M. (1977). Latitude, habitat, and hatching type for muricacean gastropods. *Nautilus*, **91**: 67–71.

Spight, T. M., Birkeland, C. E., and Lyons, A. (1974). Life histories of large and small murexes (Prosobranchia: Muricidae). *Mar. Biol.*, **24**: 229–42.

Spight, T. M. and Emlen, J. M. (1976). Clutch sizes of two marine snails with a changing food supply. *Ecology*, **57**: 1162–78.

Springer, V. G. (1982). Pacific plate biogeography, with special reference to shorefishes. *Smithsonian Contr. Zool.*, No. 367, 1–182.

Springsteen, F. J. and Leobrera, F. (1986). *Shells of the Philippines*. Carfel, Manila.

Stace, C. A. (1989). Dispersal versus vicariance—no contest. *Jour. Biogeogr.*, **16**: 201–2.

Strathmann, R. R. (1974). The spread of sibling larvae of sedentary marine invertebrates. *Amer. Nat.*, **108**: 29–44.

Strathmann, R. R. (1977). Egg size, larval development,

and juvenile size in benthic marine invertebrates. *Amer. Nat.*, **111**: 373–6.

Strathmann, R. R. (1990). Why life histories evolve differently in the sea. *Amer. Zool.*, **30**: 197–207.

Strathmann, R. R. and Chaffee, C. (1984). Constraints on egg masses. II. Effect of spacing, size, and number of eggs on ventilation of masses of embryos in jelly, adherent groups, or thin-walled capsules. *Jour. Exp. Mar. Biol. Ecol.*, **84**: 85–93.

Strathmann, R. R. and Leise, E. (1979). On feeding mechanisms and clearance rates of molluscan veligers. *Biol. Bull.*, **157**: 524–35.

Strathmann, R. R. and Vedder, K. (1977). Size and organic content of eggs of echinoderms and other invertebrates as related to developmental strategies and egg eating. *Mar. Biol.*, **39**: 305–9.

Taylor, J. P. (1975). Planktonic prosobranch veligers of Kaneohe Bay. Ph.D. dissertation, University of Hawaii.

Thorson, G. (1940*a*). Studies on the egg masses and larval development of Gastropoda from the Iranian Gulf. *Danish Scientific Investigations in Iran*, Part 2, 159–238.

Thorson, G. (1940*b*). Notes on the egg-capsules of some North-Atlantic prosobranchs of the genus *Troschelia, Chrysodomus, Volutopsis, Sipho,* and *Trophon. Vidensk. Medd. Dansk Naturh. Foren.*, **104**: 251–65.

Thorson, G. (1950). Reproductive and larval ecology of marine bottom invertebrates. *Biol. Revs.*, **25**: 1–45.

Thresher, R. E. (1987). Interoceanic and regional differences in the reproductive biology of reef fishes. *Unesco Repts. Mar. Sci.*, **46**: 219–38.

Thresher, R. E. and Brothers, E. B. (1985). Reproductive ecology and biogeography of Indo-West Pacific angelfishes (Pisces: Pomacanthidae). *Evolution*, **39**: 878–87.

Todd, C. D. (1979). Reproductive energetics of two species of dorid nudibranchs with planktotrophic and lecithotrophic larval strategies. *Mar. Biol.*, **53**: 57–68.

Trovão, H. and Rolán, E. (1986). Description of a new species for the genus *Conus* (Mollusca: Gastropoda), from the Cape Verde Islands. *Publ. Ocas. Soc. Port. Malac.*, **7**: 9–15.

Turner, R. L. and Lawrence, J. M. (1979). Volume and composition of echinoderm eggs: implications for the use of egg size in life history models. In *Reproductive ecology of marine invertebrates* (ed. S. E. Stancyk) pp. 25-40. Univ. of South Carolina Press, Columbia.

Tursch, B. and Germain, L. (1985). Studies on Olividae. I. A morphometric approach to the *Oliva* problem. *Indo-Malayan Zool.*, **2**: 331–52.

Underwood, A. J. (1991). The logic of ecological experiments: a case history from studies of the distribution of macro-algae on rocky intertidal shores. *Jour. Mar. Biol. Assn. UK*, **71**: 841–66.

Valentine, J. W. and Jablonski, D. (1983). Speciation in the shallow sea: general patterns and biogeographic controls. In *Evolution, time and space: the emergence of the biosphere* (ed. R. W. Sims, J. H. Price and P. E. S. Whalley), pp. 201–26. Academic Press, New York.

van den Berghe, E. P. and Gross, M. R. (1989). Natural selection resulting from female breeding competition in a Pacific salmon (coho: *Onchorhynchus kisutch*). *Evolution*, **43**: 125–40.

Vance, R. R. (1973). On reproductive strategies in marine benthic invertebrates. *Amer. Nat.*, **107**: 339–52.

Verduin, A. (1982). How complete are diagnoses of coiled shells of regular build? A mathematical approach. *Basteria*, **45**: 125–67.

Vermeij, G. J. (1978). *Biogeography and adaptation*. Harvard University Press, Cambridge, Massachusetts.

Vermeij, G. J. (1987). *Evolution and escalation*. Princeton University Press, Princeton, New Jersey.

Victor, B. C. (1986). Duration of the planktonic larval stage of one hundred species of Pacific and Atlantic wrasses (family Labridae). *Mar. Biol.*, **71**: 203–8.

Vink, D. L. N. and von Cosel, R. (1985). The *Conus cedonulli* complex: historical review, taxonomy, and biological observations. *Rev. Suisse Zool.*, **92**: 525–603.

Walls, J. G. [1979]. *Cone shells*. TFH Publications, Neptune City, New Jersey.

Warén, A., Carrozza, F., and Rocchini, R. (1989). *Elachisina versiliensis*, a new Mediterranean species of the family Elachisinidae (Prosobranchia, Truncatelloidea). *Boll. Malacologico*, **25**: 335–9.

Wellington, G. M. and Victor, B. C. (1989). Planktonic larval duration of one hundred species of Pacific and Atlantic damselfishes (Pomacentridae). *Mar. Biol.*, **101**: 557–67.

Werner, E. E. (1988). Size, scaling, and the evolution of complex life cycles. In *Size-structured populations: ecology and evolution* (ed. B. Ebenmann and L. Persson), pp. 60–81. Springer, Berlin.

Winkler, D. W. and Wallin, K. (1987). Offspring size and number: a life history model linking effort per offspring and total effort. *Amer. Nat.*, **129**: 708–20.

Wilson, B. R. and Gillett, K. (1971). *Australian shells*. C. E. Tuttle, Rutland, Vermont.

Zehra, I. and Perveen, R. (1990). Egg capsule structure and larval development of *Conus biliosus* (Röding, 1798) and *C. coronatus* Gmelin, 1791, from Pakistan. *Jour. Moll. Stud.*, **57**: 239–248.

INDEX

Terms preceded by * are also discussed in the individual species accounts in Chapter 3. Bold numbers denote reference to illustrations.

Acanthophora 39
Acropora 21, 29, 30, 33
asteroids, *see* echinoderms
Atlantic Ocean 7, 8, 54, 64, 68–70, 72–3, 75–6, 82, 84
Australia 14, 21, 23, 25, 27, 35, 38, 41–2, 44, 48, 69, 74

biogeographic patterns 1, 3, 57–66, 68, 70, 82–5
Bittium alternatum 79
blastula 9, 17
brooding, *see* parental care
Buccinoidea 7

Calyptraeidae 78
Cape Verde Islands 68–70, 75
Caroline Islands 13, 16, 20–4, 26–30, 33–6, 39, 40–1, 62–4
Central America 59, 70, 76
cleavage 9, 17
Clipperton Island 70
Conus
  abbreviatus 14, 36, **52**, 55, 59, **61–2**, 66, 83
  achatinus 14, 15, 65
  acutangulus 23
  ammiralis 15, **52**
  anemone 69
  araneosus 10, 15
  arenatus 15
  aristophanes 15
  aulicus 16
  aurantius 68
  balteatus 16, **61**
  bandanus 16, **62**, 82
  biliosus 5, 16, 17, 20, 55, 59, **61**, 66, 83
  canonicus 17, 18–19
  capitaneus 5, 17, 19–20
  catus 17, 20–1
  cedonulli 68
  centurio 68
  chaldaeus 10, 21, 70, 82
  cinereus 21, **52**, 59, **61**, 65
  coffeae 10, 21–2, **52**
  consors 20, 22–3, 34, **52–3**
  coronatus **4**, 5, 15, 17, **19**, 20, 23, 36, 55, 58, 59, **60**, 69, 75, 82
  cuneolus 68
  dictator 5, 23–4
  diminutus 75
  distans 24, 35
  dorreensis 69
  ebraeus 17, 21, 24, **25**, 56, 58, 59, **60**, 66, 70, 82
  eburneus 17, 24–6
  episcopus 17, 26, 53, **62**
  ermineus 68, 70
  figulinus **6**, 7, 27, 74
  flavidus 2, 27–8, 55, **62**
  floridanus 7
  frigidus 17, 28–9
  furvus **6**, 29, 47, 75
  geographus 4, 10, 29
  glans 17, 22, 29–30, **52**, 53, **62**
  guinaicus 68
  imperialis 30
  inscriptus 24
  jaspideus 5, 7, 78
  klemae 69
  leopardus 20, 30, 47–8, 56
  litteratus **6**, 30–1, **60**
  lividus 10, 17, 20, 31–2, 48, **62**
  magus 17, 20, 32–4, **52**, 56, 64–5
  mappa 68
  marmoreus 16, 20, 27, **32**, 34–5, 48, **52**, 53, **62**, 82
  miles 35, 56
  miliaris 2, 17, 35–6, 82
  moreleti 36
  mozambicus 69
  mus 68
  obscurus 36
  omaria 36, **52–3**, **62**
  papilliferus 69, 74
  pennaceus 5, 7–8, 10, 11, 20, 26, 36–7, **52**, 53, 55, **62**, 63, 66, 68, 75, 82–3
  pertusus 37
  planiliratus 23–4
  planorbis 38
  pseudaurantius 68
  pulicarius 38
  puncticulatus 68
  purpurascens 70
  quercinus 4, 8, 38–9, 55, **62**
  rattus 20, 39, **52**
  regius 68
  retifer 39–40
  scabriusculus 10, 22
  scitulus 69
  sponsalis nanus 40
  spurius 68, 75
  stercusmuscarum 40
  stramineus 10, 40, **52**
  striatellus 40, **52**
  striatus 10, 17, 41, **42**, **52**, **62**
  striolatus 41
  sumatrensis 44
  suratensis 41–2, 66, 83
  tahitensis 39
  tessulatus 42, 70
  textile 17, 18, 20, **32**, 42–3, **52**, 55–6, **62**, 73, 77
  thalassiarchus 43–4
  tinianus 69
  tulipa 4
  varius 44, **52**
  ventricosus 68, 75
  vexillum 30, 44, 47, 75
  victoriae 44
  vidua 10
  virgo 44
  vitulinus 45
Cook Islands 23, 32
coral reefs 1, 2, 6, 64, 72, 81, 83, 84
Cosmoledo Island 23–4
Crepidula 7, 78
  convexa 79
  fornicata 79
  plana 79
Cylinder 56
Cypraea 7
  annulus 15

Darioconus 56
dispersal 1–3, 6, 9, 46, 57–66, 82–5

East Pacific barrier 59, 70, 84
echinoderms 72–4, 76, 78, 80
  Asteroidea 72, 78–9, 81
  Echinoidea 39, 72–3, 76, 78–9, 81
echiurans 2
*egg capsules 5, 7–8
*egg masses 5–7, 66
eggs
  *number of 5, 8, 13–14, 46–9, 54–6, 69, 74, 75–6, 81
  *size of 5, 8–11, 14, 46–53, 62, 64, 66, 68, 70, 72–85
  *embryo, development of 3, 4, 7–10, 14, 49, 50, 62, 68, 73–5, 77–9, 81, 85
enteropneusts 2

fecundity 7, 8, 46, 49, 55, 72, 75–7
fertilization 9
Fiji 15, 16, 36, 43
fishes 2, 64, 72, 74, 76, 83–4
  angelfish 76, 84
  damselfish 84
  salmon 76
  wrasse 84
Florida 68, 79

Galapagos Islands 70
gastrula 9, 19
growth 2, 9–11, 13, 42, 54, 77, 79–80
Guam 30, 41, 64

Halimeda 43
  opuntia 33
Haminoea 73
*hatchlings 10–11, 17, 53, 71–5, 77–9
  *size of 9–11, 53, 72, 75, 77–8, 80
Hawaii 3–5, 9, 13–14, 16, 20, 24, 27, 30–21, 36–45, 50, 53, 57, 62–3, 66, 75, 79, 80, 83–4

Ilyanassa obsoleta 79
India 15, 23, 28, 36–7
Indonesia 16, 18–19, 24, 26–8, 30–3, 37, 39, 43
Isochrysis galbana 13

Johnston Island 83

Labridae 84
*larval biology 9–10
Lasaea 83
lecithotrophy 2, 11, 54, 74, 76, 79–81
Lithoconus 56
lithospheric plates 59, 62, 64–5, 83
Littorina 81
lizards 73

Madagascar 66
Malay Peninsula 66

Maldive Islands 23, 31, 36–7
Marquesas Islands 3
Marshall Islands 15, 22–4, 27, 30, 34, 38, 44, 66, 83
mating 4, 15, 29, 39
Mediterranean Sea 68, 70, 75
mesogastropods 4, 7, 77
*metamorphosis 1–4, 9–11, 13–14, 50–4, 62, 73, 78–84
Muricidae 74, 76–8
Muricoidea 7

neogastropods 4, 5, 7–9, 77
New Caledonia 19, 24, 64–5
Niue Island 21, 28
Nucella 4, 55, 74
  lamellosa 4
nurse eggs 9

Oenopota levidensis 76
*operculum 2
opisthobranchs 72–4, 79, 80, 81
*oviposition 4–5, 7, 9, 14, 48, 50–1, 55, 62, 70, 75, 78, 85

Pakistan 4–5, 16, 23, 63
Panama 73
parental care 2, 7–8, 74, 76, 81
Persian Gulf 3, 23–4, 42, 63
Petaloconchus montereyensis 76
Phaeodactylum tricornutum 13
Philippines 15, 21–3, 27, 29, 32–4, 38–41, 43–4, 57, 63–6
Pionoconus 56
*planktotrophy 2, 11, 14, 54, 69, 74, 78–81, 85
polychaetes 2, 11, 74, 80
Pomacanthidae 76, 84
Pomacentridae 84
Porites lutea 18, 42
*precompetent planktonic period (T_P) 9–10, 14, 50–1, 62–3
pre-hatching development, *see* embryo
protoconch 10, 11, 33, 50, 53–4, 68, 85

Red Sea 3
Regiconus 56
regression models 9, 46
reproductive effort 2, 7–8, 55, 76–7
reproductive energetics 14, 36, 54, 81, 85
Rhizoconus 56

Sceloporus 73
Seychelles 18, 23, 29–31, 36, 39, 44
Singapore 42
siphon 2, 29, 33
Sipuncula 72, 76
size, of mature adults 2, 3, 8, 9, 20, 48, 54–6, 75–7, 85
South Africa 69, 70, 74
spawning, *see* oviposition
Sri Lanka 29, 31, 39, 44, 66
Strombus luhuanus 4
Strongylocentrotus
  purpuratus 73
  droebachiensis 73

teleoconch 11
tentacles 2, 19, 28, 33, 35–6
Thailand 19, 21, 35, 42, 63
Thalassoma ballieui 83
T_P, *see* precompetent planktonic period
*trochophore-equivalent stage 9
Tuamotu Islands 3
Turritellidae 72, 74
turtles 76

velar lobes, *see* velum
*veliconcha 9–11, 53, 68–9, 79
*veliger 9
*velum 10, 68–9, 78–9
Venezuela 75
Vermetidae 57–8, 82, 84
Vietnam 31
vicariance 57–8, 82, 84
Virroconus 56
Volutopsis norwegicus 75

QL 430.5 .C75 K75 1994

Kohn, Alan J.

Life history and
 biogeography